# Journey
# into Mathematics
## *An Introduction to Proofs*

### Joseph J. Rotman
*University of Illinois, Urbana*

DOVER PUBLICATIONS
Garden City, New York

*Bibliographical Note*

This Dover edition, first published in 2007, is a corrected republication of the first edition of the work originally published by Prentice Hall, Inc., Upper Saddle River, New Jersey, in 1998. Readers of this book who would like to receive the solutions to the exercises may request them from the publisher at the following e-mail address: **editors@doverpublications.com**

*International Standard Book Number*

*ISBN-13: 978-0-486-45306-4*
*ISBN-10: 0-486-45306-5*

Manufactured in the United States of America
45306511    2023
www.doverpublications.com

*To Marganit.*

# Contents

## Glossary of Logic                                        **217**

## Index                                                    **233**

# Preface

Instructors have observed, when teaching junior level courses in abstract algebra, number theory, or real variables, that many students have difficulty out of proportion to the level of difficulty of the material. In an abstract algebra course introducing groups and rings, students' struggles are not affected by the changing of texts, instructors, or the order of presentation. Similarly, experimenting with courses in real variables (say, by treating only functions of a single variable instead of functions on euclidean $n$-space) offers little relief. The cause of this problem is plain when one considers the previous mathematics courses. The standard calculus sequence is presented, nowadays, to students from various disciplines who have different backgrounds, abilities, and goals, with the aim of teaching them how to differentiate, how to integrate, and how to use these techniques to solve problems. Theorems are stated but usually not proved; hypotheses of theorems are often not verified before applying the theorems (e.g., does one always check whether a given function is continuous?); definitions are given (e.g., limit and convergence) but not taken seriously. After two years of such "mathematics," is it any wonder that a junior-level student is woefully unprepared to read and do real mathematics?

There are two possible solutions to this problem. The obvious solution, revise the calculus sequence, is impractical. Many have tried; many are trying. I wish success to those still fighting the good fight, but I am pessimistic about there being a revolution in undergraduate mathematics, and I am even more pessimistic about there being such a revolution tomorrow. After all, scientists and engineers cannot afford the extra time before using calculus in their own domains, and so calculus courses are necessarily compromises between teaching the techniques of calculus and teaching an understanding of its principles.

My solution is a one semester intermediate course between calculus and the first courses in abstract algebra and real variables. This is not a new idea. There are many such "transition courses" designed to prepare students for junior-level courses, but they emphasize the elements of logic (from modus ponens and truth tables through quantifiers) and set theory (from Boolean operations through relations and functions). I find this material rather dull and uninspiring, and I imagine that this feeling is shared by most students. Of course, these things should be learned eventually; as Hermann Weyl wrote, "Logic is the hygiene that the mathematician practices to keep his ideas healthy and strong." It is cruel, however, to inflict an entire course comprised of such things on defenseless students. And it doesn't work; my unscientific observations indicate that those students who survive such tedious material do not fare any better in subsequent courses than do those who were spared. George Pólya wrote, "When introduced at the wrong time or place, good logic may be the worst enemy of good teaching." I have attached an appendix, Glossary of Logic, at the end of the book, covering much of this material. Although this section is too brief an account to qualify as a text for a standard course, it is a place where readers can look to resolve the usual questions that tend to arise.

An introductory course should contain valuable material, it must be interesting, and it must give a fairly accurate picture of what mathematics really is and what mathematicians do. One learns how to read and write proofs by reading and writing them; merely reading *about* mathematics is not an adequate substitute for actually doing mathematics. This book begins with some very elementary mathematics – induction, binomial coefficients, and polygonal areas – because, when starting out, readers must be allowed to focus on the writing and reading of proofs without the distraction of absorbing unfamiliar ideas at the same time. From the outset, complete proofs are given to serve as models for the reader. The presentation is a coherent story, with historical and etymological asides, because it is more interesting and more natural to watch a subject grow and develop. The journey continues with elementary area problems, the irrationality of $\sqrt{2}$, the Pythagorean theorem, Pythagorean triples after Diophantus, and trigonometry. The Diophantine method of finding Pythagorean triples by parametrizing the circle with rational functions is extended to finding such parametrizations of other conic sections, and this leads to a glimpse of elliptic integrals. Next, one passes to disks, proving the area and circumference formulas (due to Eudoxus and Archimedes, respectively) essentially in the classical way. This early notion of approximation is subtle, but it is digestible because one can see areas of inscribed polygons approaching the area of the disk. One notes afterward, however, that this early notion has defects. Seeing how convergence remedies defects of the classical notion gives the reader a better understanding and appreciation of the modern definition of limit. We then see why $(-1) \times (-1) = +1$, discuss the quadratic formula, complex numbers, De Moivre's theorem, the cubic formulas (Cardano's version in terms of radicals as well as Viète's trigonometric version), discriminants, and the quartic formula. The text ends with proofs of the irrationality of $e$, the irrationality of some specific values of sine and cosine, and the irrationality of $\pi$. Thus, geometry, algebra, number theory, and analysis are all intertwined. The journey travels a road from humble beginnings to a fairly sophisticated destination. I hope that students and instructors will enjoy this text, and that it will serve the several aims set forth for it.

I thank Paul Bateman, Richard Bishop, Peter Braunfeld, Everett Dade, Heini Halberstam, Carl Jockusch, Daniel Saltz, Donald Sherbert, and Kenneth Stolarsky for their excellent suggestions. I give special thanks to Philippe Tondeur, whose notes on similar material were the starting point of this text, and I also give special thanks to J.-P. Tignol, for permitting me to quote an excerpt from his wonderful book about Galois's theory of equations. I give extra special thanks to John Wetzel and Leon McCulloh who made many fine suggestions as they were teaching from a preliminary version of this text; they generously allowed me to use the ones I liked, and they did not complain in the rare cases when I did not use one. It also gives me great pleasure to thank my daughter, Ella Rose, for drawing and producing all the figures in the book.

I thank Joyce Woodworth for an excellent job of typing my manuscript in LaTeX, and I thank Adam Lewenberg for his expert help in the final stages of LaTeX typesetting. Lastly, I thank the reviewers for their good suggestions:

Linda A. Bolte, Eastern Washington University
Thomas G. Clarke, North Carolina A&T State University
Donald Nowlin, Eastern Washington University
Michael Stecher, Texas A&M University
David Walnut, George Mason University.

Joseph Rotman

## To the Reader

Histories make men wise; poets, witty; the mathematics,
subtile; natural philosophy, deep; moral, grave; logic and
rhetoric, able to contend.

*Francis Bacon*

One of the main purposes of this book is to help you learn how to read and write proofs. To further this aim, much of the early material is familiar (even at the beginning, however, there are new and interesting things) to allow you to focus on giving complete and clear proofs without distractions.

A proof is an explanation why something is true. There is a notion of formal proof, which is essentially an explanation to a machine, but we are concerned here with giving proofs to humans. Just as one does not give the same explanation to a ten-year old that one gives to an adult, one's proof, one's explanation, depends on whom one is speaking to. The audience for all of your proofs is not your instructor (who already knows the reasons!); your explanations are to be directed towards students in class, one of whom is yourself. Adequate reasons must be given to defend assertions against any possible objection; on the other hand, there is no need to explain why $3 = 3$. Try your best to say enough to persuade, and try your best not to put others to sleep by belaboring the obvious. One role of the proofs in the text is to serve as models for your own proofs. Because one becomes more sophisticated as one learns, the proofs in the text also change; certain points made explicit in the beginning are later left unsaid.

Some people think that a proof must be full of symbols, looking like ancient Egyptian hieroglyphics. Not so. Look in any mathematics book, and you will find words. Your proofs should be written in complete sentences. Of course, you may use symbols and pictures if necessary, but remember that a symbol is like a pronoun; it means nothing unless it is specified. Just as you wouldn't begin a story by saying, "He gave some of it to him there," you must not begin a proof by saying that $x = y^2$ without telling what $x$ and $y$ are (are they numbers? real? rational? integers? positive?).

Even though the context of this course is largely elementary, do not be lulled into thinking that it is an easy course with an inevitable grade of A at its conclusion. There are challenges within. If one wants the reward Bacon

xi

mentions, then there is no alternative but to do some mathematics. The journey may have some difficulties, but its goals are valuable. As Bacon says, the reward is understanding, subtlety and, we may add, pleasure.

## To the Instructor

There are several aims of this text:

to teach students how to read and write proofs;
to teach some valuable mathematics;
to show how attractive mathematics is.

Divide the course into three parts. Part I covers the first two chapters. Because students are learning the mechanics of writing proofs, one should proceed slowly. Students write proofs from the outset, using the proofs in the text, as well as the instructor's exposition, as models for their own proofs. Each student is assigned an exercise very early in the term that must be presented before the class. He or she has several days to prepare it, and prior discussion with other students and with me is encouraged. I assign no grade to the performance, the student is allowed to use any notes, and the class is allowed to heckle (I try not to make any comments until the end of the presentation). This exercise reinforces the notion that proofs are designed as explanations to the class. Regular homework assignments should be graded for presentation as well as for correctness.

Part II of the course covers Chapter 3. Some of the synthetic geometry in proving the area and circumference formulas can be done lightly, if desired. The basic idea of the chapter is to introduce convergence of sequences only after students have acquired some geometric experience with the simpler classical notion of limit.

Part III covers Chapter 4; the most important material is complex numbers and their application, via De Moivre's theorem, to real numbers.

There will not be sufficient time in most courses to cover all the material in the text. The instructor should decide on what material to omit consistent with his or her goals for the course. I have found that students (and I) enjoy the historical and etymological asides, but I do not discuss them in class unless a question arises.

*Chapter 1*

# Setting Out

## INDUCTION

So, naturalists observe, a flea
Hath smaller fleas that on him prey;
And these have smaller still to bite 'em;
And so proceed ad infinitum.

*Jonathan Swift*

There are many styles of proof, and mathematical induction is one of them. We begin by saying what mathematical induction is not. In the natural sciences, *inductive reasoning* is based on the principle that a frequently observed phenomenon will always occur. Thus, one says that the sun will rise tomorrow morning because, from the dawn of time, the sun has risen every morning. This is not a legitimate kind of proof in mathematics, for even though a phenomenon occurs frequently, it may not occur always.

Inductive reasoning is valuable in mathematics, because seeing patterns often helps in guessing what may be true. On the other hand, inductive reasoning is not adequate for proving theorems. Before we see examples, let us make sure we agree on the meaning of some standard terms.

*Definition.* An *integer* is one of $0, 1, -1, 2, -2, 3, -3, \ldots$ .

1

**Definition.** An integer $p \geq 2$ is called a *prime number*[1] if its only positive divisors are 1 and $p$. An integer $m \geq 2$ which is not prime is called *composite*.

A positive integer $m$ is composite if it has a factorization $m = ab$, where $a < m$ and $b < m$ are positive integers; the inequalities are present to eliminate the uninteresting factorization $m = m \times 1$. Notice that $a \geq 2$: since $a$ is a positive integer, the only other option is $a = 1$, which implies $b = m$ (contradicting $b < m$); similarly, $b \geq 2$.

The first few primes are 2, 3, 5, 7, 11, 13, 17, 19, 23, 29, 31, 37, 41. That this sequence never ends is proved in Exercise 2.10.

Consider the statement:

$$f(n) = n^2 - n + 41 \text{ is a prime number for every } n \geq 1$$

(this is really a whole family of statements, one for each positive integer $n$). As we evaluate $f(n)$ for $n = 1, 2, 3, 4, \ldots , 40$, we obtain the following values:

41, 43, 47, 53, 61, 71, 83, 97, 113, 131,

151, 173, 197, 223, 251, 281, 313, 347, 383, 421,

461, 503, 547, 593, 641, 691, 743, 797, 853, 911,

971, 1033, 1097, 1163, 1231, 1301, 1373, 1447, 1523, 1601.

It is tedious but not difficult (see Exercise 1.7) to prove that every one of these numbers is prime. Can we now conclude that *all* the numbers of the form $f(n)$ are prime? For example, is the next number $f(41) = 1681$ prime? The answer is no: $f(41) = 41^2 - 41 + 41 = 41^2$, which obviously factors, and hence $f(41)$ is not prime.

Here is a more spectacular example (which I first saw in an article by W. Sierpinski). A *perfect square* is an integer of the form $a^2$ for some positive integer $a$; the first few perfect squares are: 1, 4, 9, 16, 25, 36, 49. Consider the statements $S(n)$, one for each $n \geq 1$:

$$S(n) : 991n^2 + 1 \text{ is not a perfect square.}$$

It turns out that many of the statements $S(n)$ are true. In fact, the smallest number $n$ for which $S(n)$ is false is

---

[1]One reason the number 1 is not called a prime is that many theorems involving primes would otherwise be more complicated to state.

$$n = 12, 055, 735, 790, 331, 359, 447, 442, 538, 767$$
$$\approx 1.2 \times 10^{28}.$$

The original problem is a special case of **Pell's equation** (given a prime $p$, when are there integers $m$ and $n$ with $m^2 = pn^2 + 1$), and there is a way of calculating all possible solutions of it. In fact, an even more spectacular example of Pell's equation involves the prime $p = 1, 000, 099$; the smallest $n$ for which $1, 000, 099n^2 + 1$ is a perfect square has 1116 digits.) The latest scientific estimate of the age of the earth is 20 billion (20,000,000,000) years, or about $7.3 \times 10^{12}$ days, a number very much smaller than $1.2 \times 10^{28}$, let alone $10^{1115}$. If, starting on the very first day, mankind had verified statement $S(n)$ on the $n$th day, then there would be, today, as much evidence of the general truth of these statements as there is that the sun will rise tomorrow morning. And yet some statements $S(n)$ are false!

As a final example, let us consider the following statement, known as **Goldbach's Conjecture**: Every even number $m \geq 4$ is a sum of two primes. For example,

$$4 = 2 + 2$$
$$6 = 3 + 3$$
$$8 = 3 + 5$$
$$10 = 3 + 7 = 5 + 5$$
$$12 = 5 + 7$$
$$14 = 3 + 11 = 7 + 7$$
$$16 = 3 + 13 = 5 + 11$$
$$18 = 5 + 13 = 7 + 11$$
$$20 = 3 + 17 = 7 + 13$$
$$22 = 3 + 19 = 5 + 17 = 11 + 11.$$

It would be foolish to demand that all odd numbers be sums of two primes. For example, suppose that $27 = p + q$, where $p$ and $q$ are primes. If both $p$ and $q$ are odd, then their sum is even, contradicting 27 being odd. Since the only even prime is 2, we have $27 = 2 + q$, and so $q = 25$ is prime; this contradiction shows that 27 is not a sum of two primes.

No one has ever found a counterexample to Goldbach's conjecture, but neither has anyone ever proved it. At present, the conjecture has been verified for all even numbers $m \leq 10^{13}$ by H. J. J. te Riele and J.-M. Deshouillers. It has been proved by J.-R. Chen (with a simplification by P. M. Ross) that every sufficiently large even number $m$ can be written as $p + q$, where $p$ is prime and $q$ is "almost" a prime; that is, $q$ is either prime or a product of two primes. Even with all this positive evidence, however, no mathematician will say that Goldbach's conjecture must, therefore, be true for all even $m$.

We have seen what *mathematical induction* is not; let us now discuss what induction[2] is. Suppose one guesses that all the statements $S(n)$ of a certain sort are true (for example, suppose that $S(n)$ has been observed to be true for many values of $n$). Induction is a technique of proving that *all* the statements $S(n)$ are, indeed, true. For example, the reader may check that $2^n > n$ for many values of $n$, but is this inequality true for *every* value of $n$? We will prove below, using induction, that this is so.

The key idea is just this. Imagine a stairway to the sky; if its first step is white, and if the next step above a white step is also white, then all the steps of the stairway must be white. One can trace this idea back to Levi ben Gershon in 1321. There is an explicit description of induction (cited by Pascal) written by Francesco Maurolico in 1557.

Our discussion is based on the following property of positive integers (usually called the *Well Ordering Principle*).

***Least Integer Axiom***[3]. Every nonempty collection $C$ of positive integers has a smallest number in it.

Saying that $C$ is *nonempty* merely means that there is at least one integer in the collection $C$.

The Least Integer Axiom is certainly plausible. Given a nonempty collection $C$, check whether 1 is a number in $C$; if it is, then 1 is the smallest number in $C$. Otherwise, check whether 2 is a number in $C$; if it is, then 2 is the smallest number in $C$; if not, check whether 3 is a number in $C$. Continue this procedure; since there is some number in $C$, we will eventually bump into it.

---

[2]From now on, we abbreviate "mathematical induction" to "induction."

[3]There is a certain list of properties of the collection $\mathbb{N}$ of all positive integers, called axioms, from which all other properties of $\mathbb{N}$ follow. One property on this list is the Least Integer Axiom, and so it cannot be proved (unless one finds some other list of axioms characterizing $\mathbb{N}$ and derives this property from them).

The Least Integer Axiom can be restated in a more useful way.

**Theorem 1.1 (Least Criminal).** Let $S(n)$ be a family of statements, where $n$ varies over some nonempty collection of positive integers. If some of these statements are false, then there is a first false statement.

*Proof.* Let $C$ be the collection of all those positive integers $n$ for which $S(n)$ is false; by hypothesis, $C$ is nonempty. The Least Integer Axiom provides a smallest number $m$ in $C$, and $S(m)$ is the first false statement.   •

**Theorem 1.2.** Every integer $n \geq 2$ is either a prime or a product of primes.

*Proof.* Were this not so, there would be a "least criminal" $m$; that is, $m \geq 2$, $m$ is neither a prime nor a product of primes, and $m$ is the smallest such integer. Since $m$ is not a prime, it is composite: there is a factorization $m = ab$ with $a < m$ and $b < m$. Because $m$ is the least criminal, both $a$ and $b$ are "honest"; i.e., $a = pp'p'' \cdots$ for primes $p, p', p'', \ldots$, and $b = qq'q'' \cdots$ for primes $q, q', q'', \ldots$. Therefore, $m = pp'p'' \cdots qq'q'' \cdots$ is a product of (at least two) primes, a contradiction.   •

Mathematical induction is a version of Least Criminal that is usually more convenient to use.

**Theorem 1.3 (Mathematical Induction).** Let $S(n)$ be a family of statements, one for each $n \geq 1$, and suppose that:
(i) $S(1)$ is true, and
(ii) if $S(n)$ is true, then $S(n + 1)$ is true.
Then $S(n)$ is true for every $n \geq 1$.

*Proof.* It suffices to show that there are no integers $n$ for which $S(n)$ is false; that is, it suffices to show that the collection

$$C = \text{all positive integers } n \text{ for which } S(n) \text{ is false}$$

is empty.

If, on the contrary, $C$ is nonempty, then there is a first false statement, say, $S(m)$. Since $S(1)$ is true, by (i), we must have $m \geq 2$. This implies that $m - 1 \geq 1$, and so there is an $(m - 1)$st statement $S(m - 1)$ [there is no statement $S(0)$]. As $m$ is the least criminal, $m - 1$ must be honest; that is, $S(m - 1)$ is true. But (ii) says that $S(m) = S([m - 1] + 1)$ is also true, and this is a contradiction. We conclude that $C$ is empty and, hence, that all the statements are true.   •

Before we illustrate how to use mathematical induction, let us make sure we can manipulate inequalities. We recall that if two real numbers $a$ and $b$ are both positive, i.e., $a > 0$ and $b > 0$, then $ab$, $a + b$ and $1/a$ are also positive. On the other hand, the product of a positive number and a negative number is negative.

**Definition.** For any two real numbers $c$ and $d$, define

$$d < c$$

to mean that $c - d$ is positive. We write $d \leq c$ to mean either $d < c$ or $d = c$.

Notice that if $a > b$ and $b > c$, then $a > c$ [for $a - c = (a - b) + (b - c)$ is a sum of positive numbers and, hence, is itself positive]. One often abbreviates these two inequalities as $a > b > c$. The reader may check that if $a > b \geq c$, then $a > c$.

**Theorem 1.4.** Assume that $b < B$ are real numbers.
(i) If $m$ is positive, then $mb < mB$, whereas if $m$ is negative, then $mb > mB$.
(ii) For any number $N$, positive, negative, or zero, we have

$$N + b < N + B \quad \text{and} \quad N - b > N - B.$$

(iii) Let $c$ and $d$ be positive numbers. If $d < c$, then $1/d > 1/c$, and, conversely, if $1/c < 1/d$, then $c > d$.

*Proof.* (i) By hypothesis, $B - b > 0$. If $m > 0$, then the product of positive numbers being positive implies that $m(B - b) = mB - mb$ is positive; that is, $mb < mB$. If $m < 0$, then the product $m(B - b) = mB - mb$ is negative; that is, $mB < mb$.
(ii) The difference $(N + B) - (N + b)$ is positive, for it equals $B - b$. For the other inequality, $(N - b) - (N - B) = -b + B$ is positive, and, hence, $N - b > N - B$.
(iii) If $d < c$, then $c - d$ is positive. Hence, $1/d - 1/c = (c - d)/cd$ is positive, being the product of the positive numbers $c - d$ and $1/cd$ (by hypothesis, both $c$ and $d$ are positive). Therefore, $1/d > 1/c$. Conversely, if $1/c < 1/d$, then part (i) gives $d = cd(1/c) < cd(1/d) = c$; that is, $c > d$.   ●

To illustrate, since $3 < 4$, we have

$$9 \times 3 = 27 < 36 = 9 \times 4; \tag{i}$$
$$(-9) \times 3 = -27 > -36 = (-9) \times 4;$$
$$9 + 3 = 12 < 13 = 9 + 4; \tag{ii}$$
$$9 - 3 = 6 > 5 = 9 - 4;$$
$$\tfrac{1}{4} = 0.25 < 0.33 < \tfrac{1}{3}. \tag{iii}$$

It is always a good idea to see concrete examples of a theorem, for it makes the result more understandable by putting flesh on the bones of the statement. This is the first step in appreciating what a theorem means, and so it is an important habit to cultivate. Indeed, mathematics must be read with pencil and paper. If no example is given in a text, supply your own. There is an apocryphal story of a theorem so general that no particular case is known. Such a theorem would be bad mathematics.

**Theorem 1.5.** $2^n > n$ for all $n \geq 1$.

*Proof.* Regard this inequality as a sequence of statements, where the $n$th statement $S(n)$ is:

$$S(n) : 2^n > n.$$

There are two steps required for mathematical induction.
   **Base step**: The initial statement

$$S(1) : 2^1 > 1$$

is true, for $2^1 = 2 > 1$.
   **Inductive step**: If $S(n)$ is true, then $S(n+1)$ is also true; that is, if one uses the **inductive hypothesis** $S(n) : $ "$2^n > n$ is a valid inequality," then one can prove

$$S(n + 1) : \quad 2^{n+1} > n + 1.$$

First, multiply both sides of $2^n > n$ by 2; Theorem 1.4(i) gives

$$2^{n+1} = 2 \times 2^n > 2n.$$

Now $2n = n + n \geq n + 1$ (because $n \geq 1$); therefore, $2^{n+1} > 2n \geq n + 1$, and so $2^{n+1} > n + 1$, as desired.

Having verified both the base step and the inductive step, we conclude that $2^n > n$ for all $n \geq 1$. ●

Induction is plausible in the same sense that the Least Integer Axiom is plausible. Suppose that statements $S(1)$, $S(2)$, $S(3)$, . . . satisfy the hypotheses of mathematical induction. Since $S(1)$ is true, so is $S(2)$; since $S(2)$ is true, so is $S(3)$; since $S(3)$ is true, so is $S(4)$; and so forth. Induction replaces the phrase *and so forth* by the inductive step; this guarantees, for every $n$, that there is never an obstruction in the passage from a statement $S(n)$ to the next one, $S(n + 1)$.

Here are two comments before giving more applications of induction. First, one must verify both the base step and the inductive step; verification of only one of them is inadequate. For example, consider the statements $S(n)$: $n^2 = n$. The base step $S(1)$ is true, but one cannot prove the inductive step (of course, these statements are false for all $n > 1$). Another example is given by the statements $S(n)$: $n > n+1$. The next statement, $S(n+1)$, is: $n+1 > n+2$, and Theorem 1.4(ii) shows that the inductive step is true: if $n > n + 1$, then adding 1 to both sides gives $n + 1 > (n + 1) + 1$. But the base step is false (of course, all these statements are false).

Second, when first seeing induction, many people suspect that the inductive step is circular reasoning: one is using $S(n)$, and this is what one wants to prove! A closer analysis shows that this is not at all what is happening. The inductive step, by itself, does not prove that $S(n + 1)$ is true. Rather, it says that *if* $S(n)$ is true, then one can prove that $S(n+1)$ is also true. In other words, the inductive step proves that the **implication** "If $S(n)$ is true, then $S(n+1)$ is true" is correct. The truth of this implication is not the same thing as the truth of its conclusion. For example, consider the two statements: "Your grade on every exam is 100%" and "Your grade in the course is A." The implication: "If all your exams are perfect, then you will get the highest grade in the course" is true. Unfortunately, this does not say it is inevitable that your grade in the course will be A. The truth of an implication together with the truth of its hypothesis guarantee the truth of the conclusion; the truth of only the implication does not guarantee the conclusion. Our discussion above gives a mathematical example. The implication "If $n > n + 1$, then $n + 1 > n + 2$" is correct, but the conclusion "$n + 1 > n + 2$" is false. (There is a discussion of implication, from the viewpoint of truth tables, given in the Glossary at the end of the book.)

This is an appropriate time to mention the *converse* of an implication. The converse of "If $P$ is true, then $Q$ is true" is the implication "If $Q$ is true, then $P$ is true." It is possible that both an implication and its converse are true, in which case we say: "$P$ is true *if and only if* $Q$ is true." On the other hand, it is possible that an implication is true but that its converse is false. For example, the converse of the implication: "If all your exams are perfect, then you will receive the highest grade in the course" is "If you received the highest grade in the course, then all your exams were perfect." Fortunately, this converse is false. One need not be perfect to receive the grade A. According to my grading scheme, you receive the grade A in the course if and only if your exams average 90% or higher.

The next application of induction verifies a formula giving the sum of the first $n$ integers.

*Theorem 1.6.* $1 + 2 + \cdots + n = \frac{1}{2}n(n+1)$ for every $n \geq 1$.

*Proof.* The proof is by induction.

*Base step*: If $n = 1$, then the right side is $\frac{1}{2}1(1+1) = 1$ and the left side is 1, as desired.

*Inductive step*: It is always a good idea to write the $(n+1)$st statement $S(n+1)$ (so one can see what has to be proved). We must show that the sum of the first $n + 1$ integers is given by the formula:

$$1 + 2 + \cdots + n + (n+1) = \tfrac{1}{2}(n+1)(n+2).$$

Using the inductive hypothesis $S(n): 1 + 2 + \cdots + n = \frac{1}{2}n(n+1)$, we can rewrite the left side

$$[1 + 2 + \cdots + n] + (n+1) = \tfrac{1}{2}n(n+1) + (n+1),$$

and high school algebra shows that $\frac{1}{2}n(n+1)+(n+1) = \frac{1}{2}(n+1)(n+2)$. We have verified the two steps necessary for induction, and so we can conclude that the formula is true for every $n \geq 1$. •

*Example.* Here is an application of this last formula. How many pairs of positive integers $(a, b)$ are there with $a < b < 12$? If $b = 2$, then $a = 1$; if $b = 3$, then $a = 1$ or 2; if $b = 4$, then $a = 1, 2,$ or 3; $\ldots$ ; if $b = 11$, then $a = 1, 2, \ldots,$ or 10. The number of such pairs $(a, b)$ is thus $1 + 2 + \cdots + 10$, and Theorem1.6 says that this sum is equal to $\frac{1}{2}10 \times 11 = 55$.

There is a story told about the great mathematician Gauss as a boy. One of his teachers asked the students in his class to add up all the numbers from 1 to 100, thereby hoping to get some time for himself (the story assumes that no one in the school knew Theorem 1.6). But Gauss quickly volunteered that the answer is 5050. Here is what he may have done (without induction). Let $s$ denote the sum of all the numbers from 1 to 100: $s = 1+2+\cdots+99+100$. Of course, $s = 100 + 99 + \cdots + 2 + 1$. Arrange these nicely:

$$
\begin{array}{ccccccccc}
s & = & 1 & + & 2 & + & \cdots & + & 99 & + & 100 \\
s & = & 100 & + & 99 & + & \cdots & + & 2 & + & 1;
\end{array}
$$

now add the 100 columns:

$$
\begin{array}{ccccccccc}
2s & = & 101 & + & 101 & + & \cdots & + & 101 & + & 101 \\
 & = & 101 & \times & 100 & = 10,100,
\end{array}
$$

and $s = 5050$. The same argument works for any number in place of 100. Not only did Gauss give a different proof of Theorem 1.6, but he also discovered its formula. Induction is a technique of proof, but it is not a method of discovery. We displayed the formula for the sum of the first $n$ integers in Theorem 1.6, and we used induction to prove it, but we did not say how the formula was found. The formula was not discovered by induction; it arose in some other way.

*Example.* Here is a problem using both inductive reasoning and mathematical induction. We seek a formula for the sum of the first $n$ odd numbers: $1 + 3 + 5 + \cdots + (2n - 1)$. A list of the sums for $n = 1, 2, 3, 4, 5$ is 1, 4, 9, 16, 25. These are perfect squares; better, they are $1^2, 2^2, 3^2, 4^2, 5^2$. Inductive reasoning suggests the *guess*

$$S(n) : 1 + 3 + 5 + \cdots + (2n - 1) = n^2.$$

A formula has been discovered. We now use mathematical induction to prove that this guess is always true. The base step $S(1)$ has already been checked. For the inductive step, we must prove

$$S(n + 1) : [1 + 3 + \cdots + (2n - 1)] + (2n + 1) = (n + 1)^2.$$

By the inductive hypothesis $S(n) : 1+3+5+\cdots+(2n-1) = n^2$; the bracketed term on the left side is $n^2$, and so the left side equals $n^2 + (2n+1) = (n+1)^2$. By induction, $S(n)$ is true for all $n \geq 1$.

***Theorem 1.7.*** Assuming the product rule from calculus, one has

$$(x^n)' = nx^{n-1} \text{ for all } n \geq 1,$$

where $'$ denotes derivative.

*Remark.* Recall that the product rule says

$$[f(x)g(x)]' = f'(x)g(x) + f(x)g'(x).$$

*Proof.* We proceed by induction.
   *Base step.* If $n = 1$, then we are asking whether $(x)' = x^0 = 1$. Now $f'(x) = \lim_{h \to 0}(1/h)[f(x + h) - f(x)]$. When $f(x) = x$, therefore, $(x)' = \lim_{h \to 0}(1/h)[x + h - x] = \lim_{h \to 0} h/h = 1$.
   *Inductive step.* We must prove that $(x^{n+1})' = (n + 1)x^n$ using the $n$th statement $S(n) : (x^n)' = nx^{n-1}$. Since $x^{n+1} = xx^n$, the product rule and the base step give

$$(x^{n+1})' = (xx^n)' = (x)'x^n + x(x^n)'$$

$$= x^n + x(nx^{n-1}) = (n + 1)x^n.$$

We conclude that $(x^n)' = nx^{n-1}$ is true for all $n \geq 1$.   •

   We have just seen that the base step of an inductive proof need not be a triviality; sometimes it is easier to prove than the inductive step, and sometimes it is not.
   The base step of an induction may occur at an integer other than 1. For example, consider the statements

$$S(n) : 2^n > n^2.$$

Now $S(n)$ is not true for small $n$: if $n = 2$ or $4$, then there is equality, not inequality; if $n = 3$, we have $2^3 = 8 < 9 = 3^2$, an inequality in the wrong direction. However, $S(5)$ is true, for $2^5 = 32 > 25 = 5^2$.

***Theorem 1.8.*** $2^n > n^2$ is true for all $n \geq 5$.

*Proof.* We have just checked the base step $S(5)$. In proving the inductive step $2^{n+1} > (n + 1)^2$, we may assume not only the inductive hypothesis $2^n > n^2$

but also $n \geq 5$ (actually, we will need only $n \geq 3$). Multiplying $2^n > n^2$ by 2 gives

$$2^{n+1} = 2 \times 2^n > 2n^2 = n^2 + n^2 = n^2 + nn.$$

Since $n \geq 5$, we have $n \geq 3$ and so

$$nn \geq 3n = 2n + n \geq 2n + 1.$$

Therefore,

$$2^{n+1} > n^2 + 2n + 1 = (n+1)^2. \quad \bullet$$

We have seen that the base step of an induction can begin at $n = 1$ or at $n = 5$. Indeed, the base step of an induction can begin at any integer $k$; of course, the conclusion of such an induction is that the statements are true for all $n \geq k$. If there is a statement $S(0)$, one may also start an induction with base step at $n = 0$. On occasion, one finds an induction beginning at $n = -1$.[4]

One often begins an inductive proof with the phrase, "The proof is by induction on $n \geq k$." This serves to alert the reader not only that an induction is coming up and that the base step will be at $k$, but also to indicate which of several variables in a statement will be relevant to the induction. For example, consider the statement, "$(m + 1)^n > mn$." If I say that I will prove this by induction, do you expect that $m$ is fixed and that the base step $n = 1$ says $m + 1 > m$, or do you expect that $n$ is fixed and that the base step $m = 1$ says $2^n > n$? Stating that the proof is by induction on $n \geq 1$ makes one's strategy clear.

So far, we have used induction to prove some minor results; let us now use it to prove something more substantial. Observe that if $x$ and $y$ are positive real numbers, then the identity

$$(x + y)^2 = (x - y)^2 + 4xy$$

gives

$$[\tfrac{1}{2}(x + y)]^2 = xy + [\tfrac{1}{2}(x - y)]^2.$$

It follows that

$$[\tfrac{1}{2}(x + y)]^2 \geq xy,$$

---

[4] The set $\{-1, 0, 1, 2, \dots\}$ of all positive integers together with $-1$ and $0$ also satisfies the Least Integer Axiom. Let $C$ be a nonempty subset. If $-1$ is a number in $C$, then $-1$ is the smallest number in $C$; if not and if $0$ is in $C$, then $0$ is the smallest number in $C$. If neither $-1$ nor $0$ is in $C$, then $C$ is a nonempty collection of positive integers, and so the Least Integer Axiom now says that $C$ has a smallest number in it.

with the term $[\frac{1}{2}(x - y)]^2$ showing why, in general, the inequality is not an equality. If equality holds, then $[\frac{1}{2}(x - y)]^2 = 0$ and $x = y$; conversely, if $x = y$, then there is equality: $[\frac{1}{2}(x + x)]^2 = xx$, for $[\frac{1}{2}(x - x)]^2 = 0$.

Here is an application of this observation. The **hyperbolic cosine** is defined by

$$\cosh(x) = \tfrac{1}{2}(e^x + e^{-x}).$$

Since $e^x e^{-x} = 1$, it follows that

$$\cosh(x) \geq 1$$

for all $x$, with equality if and only if $e^x = e^{-x}$; that is, $\cosh(x) = 1$ if and only if $e^{2x} = 1$, so that $\cosh(x) = 1$ if and only if $x = 0$.

Given positive numbers $a_1, a_2, \ldots, a_n$, their **arithmetic mean** is defined to be their average: $A = (a_1 + a_2 + \cdots + a_n)/n$, and their **geometric mean** is defined to be $G = \sqrt[n]{a_1 a_2 \cdots a_n}$. Using these new words, we can rephrase what was shown above; the arithmetic mean of two positive numbers $a_1$ and $a_2$ is larger than their geometric mean, and equality holds precisely when $a_1 = a_2$. We are going to extend this result to many terms, but we begin with an elementary lemma followed by a normalized version of the inequality.

**Lemma 1.9.** If $0 < m < 1 < M$, then $m + M > 1 + mM$.

*Proof.* Since the product of positive numbers is positive,

$$(1 - m)(M - 1) = M - 1 - mM + m$$

is positive. Therefore, $M + m > 1 + mM$, as desired.  •

For example, there are inequalities $0 < \sin\theta < 1 < \sec\theta$ for any acute angle $\theta$, and so the lemma gives the inequality

$$\sin\theta + \sec\theta > 1 + \sin\theta \sec\theta = 1 + \tan\theta.$$

**Lemma 1.10.** If $k_1, k_2, \ldots, k_n$ are positive numbers with $k_1 k_2 \cdots k_n = 1$, then $k_1 + k_2 + \cdots + k_n \geq n$; moreover, equality holds if and only if $1 = k_1 = k_2 = \cdots = k_n$.

*Proof.* Clearly, $k_1 + k_2 + \cdots + k_n = n$ if all $k_i = 1$. Therefore, to prove both statements, it suffices to show that if $k_1 k_2 \cdots k_n = 1$ and not all $k_i = 1$, then $k_1 + k_2 + \cdots + k_n > n$. We prove this by induction on $n \geq 2$.

*Base step.* Since $k_1 k_2 = 1$ and $k_1 \neq k_2$, we may assume that $0 < k_1 < 1 < k_2$ (if both are strictly larger than 1, then $k_1 k_2 > 1$; if both are strictly smaller than 1, then $k_1 k_2 < 1$). By the lemma, $k_1 + k_2 > 1 + k_1 k_2 = 2$.

*Inductive step.* Assume that $k_1 k_2 \cdots k_{n+1} = 1$, where $k_1, k_2, \cdots, k_{n+1}$ are positive numbers. We may further assume that some $k_i < 1$: otherwise, all $k_i \geq 1$, and the present assumption that not all $k_i = 1$ gives the contradiction $k_1 k_2 \cdots k_{n+1} > 1$. For notational convenience, let $k_1 < 1$. A similar argument, with all inequalities reversed, allows us to assume that $k_{n+1} > 1$. Define $a_1 = k_1 k_{n+1}$. Note that $a_1 k_2 \cdots k_n = k_1 k_2 \cdots k_{n+1} = 1$. By the lemma,

$$k_1 + k_{n+1} > 1 + k_1 k_{n+1} = 1 + a_1,$$

so that adding $k_2 + \cdots + k_n$ to both sides gives

$$k_1 + k_2 + \cdots + k_n + k_{n+1} > 1 + a_1 + k_2 + \cdots + k_n.$$

It remains to show that $1 + a_1 + k_2 + \cdots + k_n \geq n + 1$ (for we already have an earlier strict inequality). If $a_1 = 1 = k_2 = \cdots = k_n$, then $1 + a_1 + k_2 + \cdots + k_n = n + 1$, and we are done. Otherwise, the inductive hypothesis applies and gives $a_1 + k_2 + \cdots + k_n > n$, and hence $1 + a_1 + k_2 + \cdots + k_n > n + 1$.   •

**Theorem 1.11 (Inequality of the Means).** If $a_1, a_2, \cdots, a_n$ are positive numbers, then

$$(a_1 + a_2 + \cdots + a_n)/n \geq \sqrt[n]{a_1 a_2 \cdots a_n};$$

moreover, equality holds if and only if $a_1 = a_2 = \cdots = a_n$.

*Proof.* Define $G = \sqrt[n]{a_1 a_2 \cdots a_n}$, and define $k_i = a_i/G$ for all $i$. It follows that $k_1 k_2 \cdots k_n = a_1 a_2 \cdots a_n / G^n = 1$, and so the lemma gives $k_1 + k_2 + \cdots + k_n \geq n$; that is, $a_1 + a_2 + \cdots + a_n \geq nG$, or

$$(a_1 + a_2 + \cdots + a_n)/n \geq G = \sqrt[n]{a_1 a_2 \cdots a_n}.$$

Moreover, the lemma adds that there is equality if and only if all the $k_i = 1$; that is, if and only if all the $a_i$ are equal (to $G$).   •

We shall give a geometric application of this inequality in Chapter 2: Of all the triangles with a given perimeter, the one with the greatest area must be equilateral.

The first proofs we have presented are straightforward, and it is easy to believe that we could have discovered them. The proof of the inequality of the means, however, is different; it is not so clear whether one could have discovered it without some pondering. Is there a more pedestrian proof? Probably not. Were all proofs in mathematics routine, there would be a machine that could solve any problem; press a button and wait until the machine presents its answer. But it can be proved that no such machine can ever exist. Your first reaction to this fact of life might be despair, but do not be discouraged. Mathematics is not the realm of a few "magicians"; you are not expected to compete with Archimedes, Gauss, Hilbert, and Poincaré. Each of us is inventive to some degree, and the more one learns, the more proficient one becomes. In music, we can listen and thrill to the beauty of Bach, Mozart, and Beethoven. Even though we cannot compose the sonatas for unaccompanied violin, Don Giovanni, or the late quartets, we can still sing.

The following proof of the inequality of the means is due to G. Pólya, who said that it came to him in a dream. We begin with a lemma from calculus.

**Lemma 1.12.** For all numbers $x$,

$$e^x \geq 1 + x,$$

with equality if and only if $x = 0$.

*Proof.* We consider the function $f(x) = (1 + x)/e^x = (1 + x)e^{-x}$ (notice that $f(x)$ is defined for all $x$, for the denominator $e^x$ is never 0). Now

$$f'(x) = -(1 + x)e^{-x} + e^{-x} = -xe^{-x},$$

so that $f'(x) \geq 0$ for all $x \leq 0$, $f'(0) = 0$, and $f'(x) \leq 0$ for all $x \geq 0$. Hence, $f(x)$ is an increasing function for negative $x$, it has exactly one critical point, at $x = 0$, and it is a decreasing function for positive $x$. It follows that $f(x)$ has an absolute maximum at $x = 0$; that is,

$$(1 + x)/e^x = f(x) \leq f(0) = 1 \text{ for all } x.$$

Therefore, $1 + x \leq e^x$ for all $x$, and there is equality precisely when $x = 0$.    •

Here is Pólya's proof of the inequality of the means.

**Theorem 1.13 (= Theorem 1.11).** If $a_1, a_2, \cdots, a_n$ are positive numbers, then

$$[(a_1 + a_2 + \cdots + a_n)/n]^n \geq a_1 a_2 \cdots a_n;$$

moreover, equality holds if and only if $a_1 = a_2 = \cdots = a_n$.

*Proof* (Pólya). Let us write $A = (a_1 + a_2 + \cdots + a_n)/n$ and $G = \sqrt[n]{a_1 a_2 \cdots a_n}$, so that $G^n = a_1 a_2 \cdots a_n$. We must show $A^n \geq G^n$, with equality if and only if $a_1 = a_2 = \cdots = a_n$.

For each $i$ from 1 to $n$, define $x_i = -1 + \frac{1}{A} a_i$. By the lemma,

$$e^{x_i} = e^{-1 + \frac{1}{A} a_i} \geq 1 + \left( -1 + \frac{1}{A} a_i \right) = \frac{1}{A} a_i, \tag{1}$$

and there is equality if and only if $x_i = -1 + \frac{1}{A} a_i = 0$; that is,

$$e^{x_i} = 1 + x_i \text{ if and only if } a_i = A. \tag{2}$$

We now prove the theorem. The law of exponents gives

$$\prod_{i=1}^{n} e^{-1 + \frac{1}{A} a_i} = e^{-n + \Sigma_i \frac{1}{A} a_i}.$$

But $\sum_{i=1}^{n} \frac{1}{A} a_i = (a_1 + a_2 + \cdots + a_n)/[(a_1 + a_2 + \cdots + a_n)/n] = n$. Hence, the exponent $-n + \sum_i \frac{1}{A} a_i = -n + n = 0$, and so

$$\prod_{i=1}^{n} e^{-1 + \frac{1}{A} a_i} = 1.$$

Therefore, Eq. (1) gives

$$1 = \prod_{i=1}^{n} e^{-1 + \frac{1}{A} a_i} \geq \prod_{i=1}^{n} \left( \frac{1}{A} a_i \right) = a_1 a_2 \cdots a_n / A^n = G^n / A^n; \tag{3}$$

that is, $A^n \geq G^n$.

If all the $a_i$ are equal, say, $a_i = a$ for all $i$, then $A = (a_1 + \cdots + a_n)/n = na/n = a$; that is, $a_i = A$ for all $i$. Therefore, $G^n = a_1 a_2 \cdots a_n = A^n$. Conversely, if $A^n = G^n$, then $1 = G^n/A^n$, and the inequality in Eq. (3) becomes

$$\prod_{i=1}^{n} e^{x_i} = \prod_{i=1}^{n} (1 + x_i).$$

By the lemma, $e^{x_i} \geq 1 + x_i$ for all $i$. If there is strict inequality $e^{x_k} > 1 + x_k$ for some $k$, then there is strict inequality $\prod e^{x_i} > \prod(1 + x_i)$. Therefore, $e^{x_i} = 1 + x_i$ for all $i$; by Eq. (2), this gives $a_i = A$ for all $i$. ●

This is an elegant proof, but we mere mortals must be content with more mundane ones.

There is another version of induction, usually called the *second form of induction*, that is often convenient to use.

*Definition*. The *predecessors* of an integer $n \geq 2$ are the positive integers $k$ with $k < n$, namely, $1, 2, \ldots, n - 1$.

**Theorem 1.14 (Second form of induction).** Let $S(n)$ be a family of statements, one for each $n \geq 1$, and suppose that
(i) $S(1)$ is true, and
(ii) if $S(k)$ is true for all predecessors $k$ of $n$, then $S(n)$ is itself true.
Then $S(n)$ is true for all $n \geq 1$.

*Proof*. It suffices to show that there are no integers $n$ for which $S(n)$ is false; that is, the collection

$$C = \text{all positive integers } n \text{ for which } S(n) \text{ is false}$$

is empty.

If, on the contrary, $C$ is nonempty, then there is a least criminal; that is, there is a first false statement $S(m)$. Since $S(1)$ is true, by (i), we must have $m \geq 2$. As $m$ is the *least* criminal, $k$ must be honest for all $k < m$; that is, $S(k)$ is true for all the predecessors of $m$. By (ii), $S(m)$ is true, and this is a contradiction. We conclude that $C$ is empty and, hence, that all the statements $S(n)$ are true.    •

The second form of induction can be used to give a second proof of Theorem 1.2 (as with the first form, the base step need not occur at 1).

**Theorem 1.15 (= Theorem 1.2).** Every integer $n \geq 2$ is either a prime or a product of primes.

*Proof*[5]. The base step $n = 2$ is true because 2 is a prime. Let $n > 2$. If $n$ is a prime, we are done. Otherwise, $n = ab$, where $2 \leq a < n$ and $2 \leq b < n$ (since $a$ is an integer, $1 < a$ implies $2 \leq a$). As $a$ and $b$ are predecessors of $n$, each of them is either prime or a product of primes:

$$a = pp'p'' \cdots \quad \text{and} \quad b = qq'q'' \cdots ,$$

and so $n = pp'p'' \cdots qq'q'' \cdots$ is a product of (at least two) primes.    •

---

[5]The similarity of the proofs of Theorems 1.2 and 1.15 indicates that the second form of induction is merely a variation of least criminal.

The reason the second form of induction is more convenient here is that it is more natural to use $S(a)$ and $S(b)$ than to use $S(n-1)$; indeed, it is not at all clear how to use $S(n-1)$.

Here is a notational remark. When using the second form of induction, we speak of $n$ and its predecessors, not $n+1$ and its predecessors. If one wants to compare the two forms of induction, one could say that the first form uses $S(n-1)$ to prove $S(n)$, whereas the second form uses any or all of the earlier statements $S(1), S(2), \ldots, S(n-1)$, to prove $S(n)$.

The next result says that one can always factor out a largest power of 2 from any integer.

**Theorem 1.16.** Every positive integer $n$ has a factorization $n = 2^k m$, where $k \geq 0$ and $m \geq 1$ is odd.

*Proof.* We use the second form of induction on $n \geq 1$.

*Base step*: If $n = 1$, take $k = 0$ and $m = 1$.

*Inductive step*: If $n \geq 1$, then $n$ is either odd or even. If $n$ is odd, then take $k = 0$ and $m = n$. If $n$ is even, then $n = 2b$. Because $b < n$, it is a predecessor of $n$, and so the inductive hypothesis allows us to assume $S(b) : b = 2^\ell m$, where $\ell \geq 0$ and $m$ is odd. The desired factorization is $n = 2b = 2^{\ell+1} m$. •

**Definition.** The **Fibonacci sequence** $F_0, F_1, F_2, \ldots$ is defined as follows:

$$F_0 = 0, \quad F_1 = 1, \quad \text{and} \quad F_n = F_{n-1} + F_{n-2} \quad \text{for all } n \geq 2.$$

Thus, the sequence begins: $0, 1, 1, 2, 3, 5, 8, 13, \ldots$ .

**Theorem 1.17.** If $F_n$ denotes the $n$th term of the Fibonacci sequence, then

$$F_n = \tfrac{1}{\sqrt{5}}(\alpha^n - \beta^n)$$

for all $n \geq 0$, where $\alpha = \tfrac{1}{2}(1 + \sqrt{5})$ and $\beta = \tfrac{1}{2}(1 - \sqrt{5})$.

*Remark.* The number $\alpha = \tfrac{1}{2}(1 + \sqrt{5})$ is called the **golden ratio**. The ancient Greeks called a rectangular figure most pleasing if its edges $a$ and $b$ were in the proportion $a : b = b : a + b$. It follows that $b^2 = a(a + b)$, so that $b^2 - ab - a^2 = 0$, and the quadratic formula gives $b = \tfrac{1}{2}(a \pm \sqrt{a^2 + 4a^2}) = a\tfrac{1}{2}(1 \pm \sqrt{5})$. Therefore,

$$b/a = \alpha \quad \text{or} \quad b/a = \beta.$$

*Proof.* We are going to use the second form of induction [the second form is the appropriate induction here, for the equation $F_n = F_{n-1} + F_{n-2}$ suggests that proving $S(n)$ will involve not only $S(n-1)$ but $S(n-2)$ as well].

*Base step.* The formula is true for $n = 0$ : $\frac{1}{\sqrt{5}}(\alpha^0 - \beta^0) = 0 = F_0$ and

$$\frac{1}{\sqrt{5}}(\alpha^1 - \beta^1) = \frac{1}{\sqrt{5}}(\alpha - \beta)$$

$$= \frac{1}{\sqrt{5}}\left[\frac{1}{2}(1 + \sqrt{5}) - \frac{1}{2}(1 - \sqrt{5})\right]$$

$$= \frac{1}{\sqrt{5}}(\sqrt{5}) = 1 = F_1$$

(we have mentioned both $n = 0$ and $n = 1$ because the inductive step will use two predecessors).

*Inductive step.* If $n \geq 2$, then

$$F_n = F_{n-1} + F_{n-2}$$
$$= \frac{1}{\sqrt{5}}(\alpha^{n-1} - \beta^{n-1}) + \frac{1}{\sqrt{5}}(\alpha^{n-2} - \beta^{n-2})$$

$$= \frac{1}{\sqrt{5}}\left[(\alpha^{n-1} + \alpha^{n-2}) - (\beta^{n-1} + \beta^{n-2})\right]$$

$$= \frac{1}{\sqrt{5}}\left[\alpha^{n-2}(\alpha + 1) - \beta^{n-2}(\beta + 1)\right]$$

$$= \frac{1}{\sqrt{5}}(\alpha^n - \beta^n),$$

because the numbers $\alpha$ and $\beta$ are the roots of the quadratic equation

$$x^2 = x + 1,$$

so that $\alpha + 1 = \alpha^2$ and $\beta + 1 = \beta^2$. •

*Remark.* It is curious that the integers $F_n$ are expressed in terms of the irrational number $\sqrt{5}$. An analogous phenomenon will be seen later: there are formulas that express real numbers in terms of complex numbers.

One can also use induction to give definitions. For example, we can define **n factorial**, denoted by $n!$, by induction on $n \geq 0$. Define $0! = 1$, and if $n!$ is known, then define

$$(n + 1)! = n!(n + 1).$$

The reason for defining $0! = 1$ will be apparent in the next section.

## Exercises

**1.1.** Find a formula for $1 + \sum_{j=1}^{n} j!\,j$, and use mathematical induction to prove that your formula is correct.

(*Remark.* This exercise illustrates the two types of induction described at the beginning of the chapter: your guess uses inductive reasoning, while its proof using base and inductive steps is mathematical induction.)

**1.2.** If $r \neq 1$, prove, for all $n \geq 1$, that

$$1 + r + r^2 + r^3 + \cdots + r^{n-1} = \frac{r^n - 1}{r - 1}.$$

**1.3.** Show, for all $n \geq 1$, that $10^n$ leaves remainder 1 after dividing by 9. (Hint: Prove $10^n = 9q_n + 1$ for some integer $q_n$.)

**1.4.** If $a \leq b$ are positive numbers, prove that $a^n \leq b^n$ for all integers $n \geq 0$.

**1.5.** (i) Prove that $1^2 + 2^2 + \cdots + n^2 = \frac{1}{6}n(n+1)(2n+1)$.
(ii) Prove that $1^3 + 2^3 + \cdots + n^3 = (1+2+\cdots+n)^2$. (Hint: Use Theorem 1.6.)
(iii) Prove that $1^4 + 2^4 + \cdots + n^4 = \frac{1}{5}n^5 + \frac{1}{2}n^4 + \frac{1}{3}n^3 - \frac{1}{30}n$.

**1.6.** (i) Find a formula for $a_n = 1^3 + 3^3 + 5^3 + \cdots + (2n-1)^3$, and then prove that your guess is correct using induction.
(ii) Give a second proof of part (i) based on Exercise 1.5(ii) and the following observation: If $b_m = 1^3 + 2^3 + 3^3 + \cdots + m^3$, then

$$b_{2n} = a_n + [2^3 + 4^3 + \cdots + (2n)^3]$$
$$= a_n + 8[1^3 + 2^3 + \cdots + n^3] = a_n + 8b_n.$$

**1.7.** (i) Prove that if $n = ab$, where $n$, $a$, and $b$ are positive integers, then either $a \leq \sqrt{n}$ or $b \leq \sqrt{n}$.
(ii) Prove that if $n$ is composite, then it has a prime factor $p$ with $p \leq \sqrt{n}$. Conclude that if $n \geq 2$ has no prime factors $\leq \sqrt{n}$, then $n$ is a prime.
(iii) If $f(n) = n^2 - n + 41$, use part (ii) to show that $f(10)$, $f(20)$, $f(30)$, and $f(40)$ are prime. (If each student in a class checks that two or three values of $f(n)$ are prime, then the class will have shown that $f(n)$ is prime for all $n \leq 40$.)

*Remark.* It is now simple to check that 991 is a prime, but checking that 1,000,099 is a prime is a longer enterprise, for its square root is a bit over 1000.

**1.8.** Prove, for all $n \geq 0$, that $(1 + x)^n \geq 1 + nx$ if $1 + x > 0$.

**1.9.** (i) Prove that $2^n > n^3$ for all $n \geq 10$.
(ii) Prove that $2^n > n^4$ for all $n \geq 17$.

**1.10.** Let $g_1(x), \ldots, g_n(x)$ be differentiable functions. If $f(x) = g_1(x) \cdots g_n(x)$, prove that its derivative is

$$f'(x) = \sum_{i=1}^{n} f(x) g_i'(x) / g_i(x).$$

**1.11.** Prove that every positive integer $a$ has a factorization $a = 3^k m$, where $k \geq 0$ and $m$ is not a multiple of 3. (Hint: Adapt the proof of Theorem 1.16.)

*Remark.* This last exercise illustrates another reason for knowing proofs. The solution of Exercise 1.11 does not follow from the statement of Theorem 1.16, but a solution can be obtained by modifying the proof of that theorem.

**1.12.** Prove that $2^n < n!$ for all $n \geq 4$.

**1.13.** Prove that $F_n < 2^n$ for all $n \geq 0$, where $F_0, F_1, F_2, \ldots$ is the Fibonacci sequence.

**1.14.** For every acute angle $\theta$, i.e., $0° < \theta < 90°$, prove that

$$\sin\theta + \cot\theta + \sec\theta \geq 3.$$

(Hint: Use the inequality of the means or Lemma 1.10.)

**1.15.** Prove that if $a_1, a_2, \ldots, a_n$ are positive numbers, then

$$(a_1 + a_2 + \cdots + a_n)(1/a_1 + 1/a_2 + \cdots + 1/a_n) \geq n^2.$$

**1.16.** For every $n \geq 2$, prove that there are $n$ consecutive composite numbers; that is, there is some integer $b$ such that $b + 1, b + 2, \ldots, b + n$ are all composite. (Hint: If $2 \leq a \leq n + 1$, then $a$ is a divisor of $(n + 1)! + a$.)

## BINOMIAL COEFFICIENTS

Let no one say that I have said nothing new ... the arrangement of the
subject is new. When we play tennis, we both play with the same ball,
but one of us places it better.

*B. Pascal*

Consider the formulas for powers $(1+x)^n$ of the binomial $1+x$ :

$$(1+x)^0 = 1$$

$$(1+x)^1 = 1+1x$$

$$(1+x)^2 = 1+2x+1x^2$$

$$(1+x)^3 = 1+3x+3x^2+1x^3$$

$$(1+x)^4 = 1+4x+6x^2+4x^3+1x^4.$$

Is there a pattern in the coefficients in these formulas?  Figure 1.1,
called *Pascal's triangle* [after B. Pascal (1623–1662)], shows the first few
coefficients.

```
               1
             1   1
           1   2   1
         1   3   3   1
       1   4   6   4   1
     1   5  10  10   5   1
   1   6  15  20  15   6   1
```

Figure 1.1

Figure 1.2 is a Chinese illustration made in the year 1303, which shows that Pascal's triangle had been recognized long before Pascal.

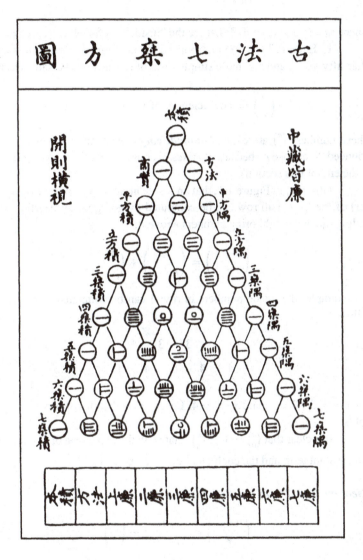

Figure 1.2

The expansion of $(1 + x)^n$ is an expression of the form

$$c_0 + c_1 x + c_2 x^2 + \cdots + c_n x^n,$$

where $c_0 = 1$ and $c_n = 1$. What are the "inside" coefficients $c_1, \ldots, c_{n-1}$?

L. Euler (1707–1783) introduced the notation $\left(\frac{n}{r}\right)$, which lost its bar after fifty years, and this more simple form of it is now universally accepted:

$$\binom{n}{r} = \text{coefficient } c_r \text{ of } x^r \text{ in } (1 + x)^n.$$

These numbers $\binom{n}{r}$ are called **binomial coefficients**; the number $\binom{n}{r}$ is pronounced "$n$ choose $r$" because it arises in counting problems (as we shall see at the end of this section).

Observe, in Figure 1.1, that an inside number (i.e., not a 1 on the border) of the $(n + 1)$th row can be computed by going up to the $n$th row and adding the two neighboring numbers above it:

$$\binom{n+1}{r} = \binom{n}{r-1} + \binom{n}{r}.$$

For example, the inside numbers in row 4 can be computed from row 3 as follows:

$$\begin{array}{ccccccc} & 1 & & 3 & & 3 & & 1 \\ 1 & & 4 & & 6 & & 4 & & 1 \end{array}$$

$1+3 = 4$, $3+3 = 6$, and $3+1 = 4$. Let us prove that this observation always holds.

It is clear that $\binom{n}{0} = 1 = \binom{n}{n}$; that is, in the expansion of $(1+x)^n$, both the constant term and the coefficient of $x^n$ are equal to 1.

**Theorem 1.18.** For all $r$ with $0 < r < n + 1$,

$$\binom{n+1}{r} = \binom{n}{r-1} + \binom{n}{r}.$$

*Proof.* We must show that if

$$(1 + x)^n = c_0 + c_1 x + c_2 x^2 + \cdots + c_n x^n,$$

then the coefficient of $x^r$ in $(1+x)^{n+1}$ is $c_{r-1} + c_r$.

$$\begin{aligned}
(1+x)^{n+1} &= (1+x)(1+x)^n \\
&= (1+x)^n + x(1+x)^n \\
&= (c_0 + c_1 x + c_2 x^2 + \cdots + c_n x^n) + \\
&\quad x(c_0 + c_1 x + c_2 x^2 + \cdots + c_n x^n) \\
&= (c_0 + c_1 x + c_2 x^2 + \cdots + c_n x^n) + \\
&\quad c_0 x + c_1 x^2 + c_2 x^3 + \cdots + c_n x^{n+1} \\
&= 1 + (c_0 + c_1)x + (c_1 + c_2)x^2 + (c_2 + c_3)x^3 + \cdots,
\end{aligned}$$

because $c_0 = 1$. Thus, the coefficient of $x^r$ in $(1+x)^{n+1}$ is $c_{r-1} + c_r$.    •

We shall see how the formula in the next theorem arises when we discuss some counting problems.

**Theorem 1.19 (Pascal).** For all $n \geq 0$ and all $r$ with $0 \leq r \leq n$,

$$\binom{n}{r} = \frac{n!}{r!(n-r)!}.$$

*Proof.* We prove the theorem by induction on $n \geq 0$.

The base step $n = 0$ is easy: by definition, $\binom{0}{0} = 1$, while $0!/0!0! = 1$ as well. To prove the inductive step, we must show

$$\binom{n+1}{r} = \frac{(n+1)!}{r!(n+1-r)!}.$$

By Theorem 1.18,

$$\binom{n+1}{r} = \binom{n}{r-1} + \binom{n}{r}$$

$$= \frac{n!}{(r-1)!(n-r+1)!} + \frac{n!}{r!(n-r)!}$$

$$= \frac{n!}{(r-1)!(n-r)!} \cdot \left( \frac{1}{n-r+1} + \frac{1}{r} \right)$$

$$= \frac{n!}{(r-1)!(n-r)!} \cdot \left( \frac{r+n-r+1}{r(n-r+1)} \right)$$

$$= \frac{n!}{(r-1)!(n-r)!} \cdot \left( \frac{n+1}{r(n-r+1)} \right)$$

$$= \frac{(n+1)!}{r!(n+1-r)!}. \quad \bullet$$

One defines $0! = 1$ to make formulas like this one more simple; without this convention, there would have to be an extra statement giving the formula in the special cases $r = 0$ and $r = n$. Moreover, the base step $n = 0$ is simpler than the base step $n = 1$.

*Corollary 1.20.* For any number $x$ and for all $n \geq 0$,

$$(1+x)^n = \sum_{r=0}^{n} \binom{n}{r} x^r = \sum_{r=0}^{n} \frac{n!}{r!(n-r)!} x^r.$$

*Proof.* The binomial coefficients have been defined as the numbers $c_r$, where

$$(1+x)^n = c_0 + c_1 x + c_2 x^2 + \cdots + c_n x^n.$$

Since $c_r = \binom{n}{r}$, we have

$$(1+x)^n = \binom{n}{0} + \binom{n}{1}x + \binom{n}{2}x^2 + \cdots + \binom{n}{r}x^r + \cdots + \binom{n}{n}x^n,$$

and the formula in Pascal's theorem now gives the result.   •

*Corollary 1.21 (Binomial Theorem).* For all numbers $a$ and $b$ and for all integers $n \geq 0$,

$$(a+b)^n = \sum_{r=0}^{n} \binom{n}{r} b^r a^{n-r} = \sum_{i+j=n} \binom{n}{i} b^i a^j.$$

*Proof.* The result is trivially true when $a = 0$ (we agree upon the notation $0^0 = 1$). If $a \neq 0$, set $x = b/a$ in Corollary 1.20, and observe that

$$\left(1 + \frac{b}{a}\right)^n = \left(\frac{a+b}{a}\right)^n = \frac{(a+b)^n}{a^n}.$$

Thus, $(a+b)^n$ is obtained by multiplying the expression for $(1 + b/a)^n$ by $a^n$.   •

We have used a second convention: $0^0 = 1$, and we are using it for the same reason we use $0! = 1$: it simplifies the writing of formulas.

Here is a combinatorial interpretation of the binomial coefficients. Given a set $X$, an *r-subset* is a subset of $X$ with exactly $r$ elements. If $X$ has $n$ elements, denote the number of its $r$-subsets by

$$[n, r].$$

We compute $[n, r]$ by considering a related question. Given an "alphabet" with $n$ (distinct) letters and a number $r$ with $1 \leq r \leq n$, an $r$-letter *word* is a sequence of $r$ of the letters with no repetitions (and with no regard to whether the "word" actually occurs in some dictionary). For example, the 2-letter words on the alphabet $a, b, c$ are

$$ab, ba, ac, ca, bc, cb$$

(note that $aa, bb, cc$ are not on this list). How many $r$-letter words are there on an alphabet with $n$ letters? We count the number of words in two ways.

(I) There are $n$ choices for the first letter; having chosen the first letter, there are now only $n-1$ choices for the second letter, for no letter is repeated; having chosen the first two letters, there are only $n - 2$ choices for the third letter, and so forth. Thus, the number of $r$-letter words is

$$n(n - 1)(n - 2) \cdots (n - [r - 1]) = n(n - 1)(n - 2) \cdots (n - r + 1).$$

Note the special case $n = r$: the number of $n$-letter words on $n$ letters is $n!$.

(II) Here is a second way to count these words. First choose an $r$-subset of the alphabet (consisting of $r$ letters); there are $[n, r]$ ways to do this, for this is exactly what the symbol $[n, r]$ means. For each chosen $r$-subset, there are $r!$ ways to arrange the $r$ letters in it (this is the special case of (I) when $n = r$). The number of $r$-letter words is thus

$$r![n, r].$$

We conclude that

$$r![n, r] = n(n - 1)(n - 2) \cdots (n - r + 1).$$

Therefore,

$$
\begin{aligned}
[n, r] &= \frac{n(n - 1)(n - 2) \cdots (n - r + 1)}{r!} \\
&= \frac{n(n - 1)(n - 2) \cdots (n - r + 1)}{r!} \cdot \frac{(n - r)!}{(n - r)!} \\
&= \frac{n!}{r!(n - r)!} = \binom{n}{r}
\end{aligned}
$$

(the last equation by Pascal's theorem). This fact is the reason one often pronounces the binomial coefficient $\binom{n}{r}$ as $n$ choose $r$.

As an example, how many ways are there to choose 2 hats from a closet containing 14 different hats? (One of my friends does not like the phrasing of this question. After all, one can choose 2 hats with one's left hand, with one's right hand, with one's teeth, . . . ; but I continue the evil tradition.) The answer is $\binom{14}{2}$, and Pascal's formula allows us to compute this as $14 \times 13/2 = 91$.

Our first interpretation of the binomial coefficients $\binom{n}{r}$ was *algebraic*; that is, as Pascal's formula in terms of factorials; our second interpretation is *combinatorial*; that is, as $n$ choose $r$. Quite often, each interpretation can be used to prove a desired result. For example, let us prove Theorem 1.18 combinatorially. Let $X$ be a collection of $n + 1$ balls, and let us color one ball red and the other $n$ balls blue. Now $\binom{n+1}{r}$ is the number of $r$-subsets $S$ of $X$. There are two possibilities for an $r$-subset $S$: either it contains the red ball or it does not. If $S$ does contain the red ball, then $S$ consists of the red ball and $r - 1$ blue balls, and so the number of such $S$ is the same as the number of $(r - 1)$-subsets of the blue balls, namely, $\binom{n}{r-1}$. The other possibility is that $S$ consists completely of blue balls; since there are $n$ blue balls, there are $\binom{n}{r}$ such $r$-subsets. Therefore, $\binom{n+1}{r} = \binom{n}{r-1} + \binom{n}{r}$, as desired.

## Exercises

**1.17.** Show that the binomial coefficients are "symmetric": if $0 \leq r \leq n$, then

$$\binom{n}{r} = \binom{n}{n-r}.$$

**1.18.** Show, for every $n$, that the sum of the binomial coefficients is $2^n$:

$$\binom{n}{0} + \binom{n}{1} + \binom{n}{2} + \cdots + \binom{n}{n} = 2^n.$$

**1.19.** (i) Show, for every $n > 0$, that the "alternating sum" of the binomial coefficients is zero:

$$\binom{n}{0} - \binom{n}{1} + \binom{n}{2} - \cdots \pm \binom{n}{n} = 0.$$

(ii) Use part (i) to prove, for a given $n$, that the sum of all the binomial coefficients $\binom{n}{r}$ with $r$ even is equal to the sum of all those $\binom{n}{r}$ with $r$ odd.

**1.20.** What is the coefficient of $x^{16}$ in $(1 + x)^{20}$?

**1.21.** How many ways are there to choose 4 colors from a palette containing 20 different paints?

**1.22.** Prove that a set $X$ with $n$ elements has exactly $2^n$ subsets. Can you give more than one proof of this?

**1.23.** A weekly lottery asks you to select 5 numbers between 1 and 45. At the week's end, 5 such numbers are drawn at random, and you win the jackpot if all your numbers match, in some order, the drawn numbers. How many selections of 5 numbers are there?                    Answer: $1, 221, 759$.

**1.24.** Assume that *term-by-term differentiation* of power series is valid: if

$$f(x) = \sum_{k \geq 0} a_k x^k = a_0 + a_1 x + a_2 x^2 + \cdots + a_n x^n + \cdots ,$$

then the power series for its derivative $f'(x)$ is

$$f'(x) = \sum_{k \geq 1} k a_k x^{k-1} = a_1 + 2a_2 x + 3a_3 x^2 + \cdots + n a_n x^{n-1} + \cdots .$$

(i) Prove that $f(0) = a_0$.
(ii) Prove, for all $n \geq 0$, that the $n$th derivative

$$f^{(n)}(x) = \sum_{k \geq n} k(k-1)(k-2) \cdots (k-n+1) a_k x^{k-n}.$$

$[f^{(0)}(x)$ is defined to be $f(x)].^6$ Conclude, for all $n \geq 0$, that

$$a_n = \frac{f^{(n)}(0)}{n!}.$$

**1.25.** (*Leibniz*) A real-valued function $f(x)$ is called a $C^\infty$-*function* if it has an $n$th derivative $f^{(n)}(x)$ for every $n \geq 0 [f^{(0)}(x)$ is defined to be $f(x)]$. Prove that if $f(x)$ and $g(x)$ are $C^\infty$-functions, then

$$(f(x)g(x))^{(n)} = \sum_{k=0}^{n} \binom{n}{k} f^{(k)}(x) g^{(n-k)}(x) \text{ for all } n \geq 0.$$

**1.26.** Prove, for all $n \geq 1$ and for all $r > 1$, that

$$\binom{n}{0} + \binom{n+1}{1} + \binom{n+2}{2} + \cdots + \binom{n+r}{r} = \binom{n+r+1}{r}.$$

(Hint: Use induction on $r \geq 2$.)

---

$^6$Our previous notational conventions, $0! = 1$ and $0^0 = 1$, were also introduced to simplify formulas, and they are as harmless as this one.

*Chapter 2*

# Things Pythagorean

### AREA

Let no one ignorant of geometry enter my door.

*Plato*

One of the earliest ideas of geometry is the computation of the area of geometric figures. Let us start with the simplest figure; a rectangle with sides of lengths $a$ and $b$ has area $A = ba$.

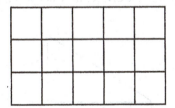

Figure 2.1

For instance, if the base is of length $b = 5$ and the height is $h = 3$, then the area is 15 units of area. If we cut the rectangle in half, we obtain two congruent right triangles:

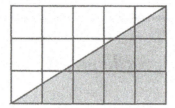

Figure 2.2

The area $A$ of a right triangle is, thus, $\frac{1}{2}$ base $\times$ height.

The area $A$ of an arbitrary triangle of base $b$ and height $h$, as you recall, is also $A = $ area $= \frac{1}{2}bh$. Observe that this formula is really three formulas in

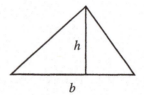

Figure 2.3

one, because each of the three sides of a triangle qualifies as a "base" of the triangle; the corresponding height is the length of the altitude to that side.

Let us prove the area formula by using the formula for a right triangle. If, as in Figure 2.4, the altitude divides the triangle into two right triangles of

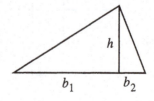

Figure 2.4

height $h$ having bases $b_1$ and $b_2$, respectively (so that $b = b_1 + b_2$), then the area $A$ is

$$A = \tfrac{1}{2}b_1h + \tfrac{1}{2}b_2h = \tfrac{1}{2}(b_1 + b_2)h = \tfrac{1}{2}bh.$$

Suppose now that the altitude does not divide the triangle into two triangles, but that we have Figure 2.5; that is, the possibilities are either that $S$ lies

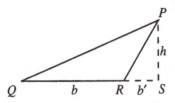

Figure 2.5

between $Q$ and $R$ or that it does not lie between them. Let $b'$ be the length of $RS$, so that $b + b'$ is the length of $QS$. Since $\triangle QSP$ and $\triangle RSP$ are right triangles, their areas are $\frac{1}{2}(b+b')h$ and $\frac{1}{2}b'h$, respectively. The area of $\triangle PQR$ is, thus, $\frac{1}{2}(b + b')h - \frac{1}{2}b'h = \frac{1}{2}(b + b' - b')h = \frac{1}{2}bh$. Having considered all the cases, we can now declare the result.

***Theorem 2.1.*** The area $A$ of any triangle with base $b$ and height $h$ is given by

$$A = \tfrac{1}{2}bh.$$

Let $PA$ and $QR$ be parallel, and consider the triangles $\triangle PQR$ and $\triangle AQR$ in Figure 2.6a. Since they have the same base and the same height, they have the same area, by Theorem 2.1. That is, if we hold the base fixed and slide the top along a horizontal line, then the area remains the same.

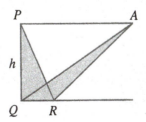

Figure 2.6a

This invariance of area after "sliding" was already recognized by Euclid, one of whose theorems is: "Triangles drawn to a point are to each other as their bases."

Euclid's theorem is illustrated in Figure 2.6$b$:

$$\frac{\text{area}(\triangle ABC)}{\text{area}(\triangle ADE)} = \frac{|BC|}{|DE|}.$$

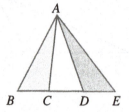

Figure 2.6$b$

Exercise 2.3 below shows that the two triangles in Figure 2.6$a$ have horizontal cross sections of equal length. *Cavalieri's principle*, named after B. Cavalieri (ca. 1598–1647), extends the invariance of area after "sliding" to more general figures.

*Cavalieri's Principle*: Two figures with horizontal cross sections of equal length have the same area.

The next result shows that Cavalieri's principle applies to parallelograms: holding the base fixed and sliding the top does not change the area.

*Theorem 2.2*. The area $A$ of a parallelogram with height $h$ and base $b$ is given by $A = hb$.

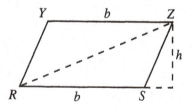

Figure 2.7

*Proof.* Divide the parallelogram with height $h$ and base $b$ into two triangles, as in Figure 2.7. Now $\triangle ZRS$ has base $RS$ of length $b$, height $h$, and area $\frac{1}{2}hb$; similarly, $\triangle YZR$ has base $YZ$ of length $b$, height $h$, and area $\frac{1}{2}hb$. We conclude that the parallelogram has area $\frac{1}{2}hb + \frac{1}{2}hb = hb$.    •

Another way to see that the area of a parallelogram is $hb$ is indicated in Exercise 2.1. As the proof of Theorem 2.1, the proof suggested in this exercise also involves two cases, the second of which is often overlooked.

We sketch two proofs of Cavalieri's principle using calculus. The first proof involves the computation of the area of a region $R$ by double integrals.

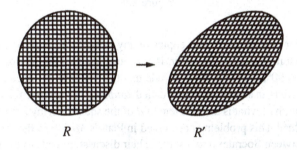

$R$                    $R'$

Figure 2.8

The basic idea is to approximate this area by superimposing a grid of squares on it. If $R$ "slides" into a region $R'$ (that is, if $R$ and $R'$ have horizontal cross sections of equal length), then a grid of squares approximating the area of $R$ slides into a grid of parallelograms approximating the area of $R'$. Because each square slides into a parallelogram of the same area, by Theorem 2.2, one can conclude that $R$ and $R'$ have the same area.

A second proof involves ordinary integrals. If $R$ is a region in the plane lying in the strip bounded by the lines with equations $y = a$ and $y = b$, and if $r(y)$ is the length of the cross section cutting $R$ at height $y$, then

$$\text{area}(R) = \int_a^b r(y)\,dy.$$

If $S$ is another region in this strip having cross sections $s(y)$, and if $r(y) = s(y)$ for all $y$, then

$$\text{area}(S) = \int_a^b s(y)\,dy = \int_a^b r(y)\,dy = \text{area}(R).$$

The following elementary observation will be needed soon. A parallelogram having all 4 sides of equal length, called a *rhombus*, need not be a

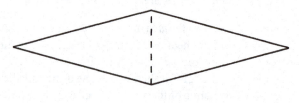

Figure 2.9

square. For example, take two copies of any isosceles triangle and glue them together along the base. However, if one can show that an interior angle of a rhombus is 90°, then one can conclude that the rhombus is, indeed, a square.

Here is an ancient problem, called *doubling the square*. Given a square of side length $a$, what is the side length $d$ of the square having double its area, namely, $2a^2$? This problem is discussed in Plato's *Menon* in the form of a dialogue between Socrates and a slave. Their discussion leads to the geometric construction in Figure 2.10.

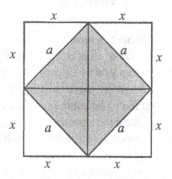

Figure 2.10

**Theorem 2.3**. The length $d$ of the diagonal of a square having side length $a$ satisfies the equation

$$d^2 = 2a^2.$$

*Proof.* In Figure 2.10, the large region is a square with all sides of length $d$, where $d = 2x$. The shaded region is a rhombus with all sides of length $a$; it is a square because the base angles of the right triangles are each 45°, and hence the interior angles of the rhombus are each $45° + 45° = 90°$.

We compute the area of the shaded square in two ways. On the one hand, it has area $a^2$, for the length of a side is $a$. On the other hand, this square is divided into 4 right triangles, each of area $\frac{1}{2}x^2$. Therefore, $a^2 = 4 \times \frac{1}{2}x^2 = 2x^2$, and

$$2a^2 = 4x^2 = (2x)^2 = d^2. \quad \bullet$$

We remark that there is a quick proof of the theorem using the Pythagorean theorem, for $d^2 = a^2 + a^2 = 2a^2$. Indeed, we have just proved the special case of the Pythagorean theorem for isosceles right triangles. We will prove the Pythagorean theorem in the next section.

**Corollary 2.4.** Let $C$ be a circle, let $S$ be a circumscribed square, and let $T$ be an inscribed square. Then

$$\text{area}(S) = 2\,\text{area}(T).$$

*Proof.* Using the notation in Figure 2.10 (imagine a circle inscribed in the big square, circumscribing the shaded square), the diameter of $C$ is equal to $d = 2x$. By the theorem,

$$\text{area}(S) = d^2 = 2a^2 = 2\,\text{area}(T). \quad \bullet$$

Before continuing this discussion, we remind you of some familiar terminology.

**Definition.** A *divisor* of an integer $m$ is an integer $d$ for which

$$m = dq,$$

where $q$ is an integer.

Recall that if $m$ and $d$ are positive integers, then long division gives integers $q$ and $r$ with

$$\frac{m}{d} = q + \frac{r}{d}$$

and $r/d < 1$. Clearing denominators, this equation involving fractions gives an equation of integers

$$m = dq + r,$$

where the **quotient** $q$ is an integer $\geq 0$ and the **remainder** $r$ is one of $0, 1, 2,$ $\ldots, d-1$. Of course, $d$ is a divisor of $m$ precisely when the remainder $r = 0$.

For example, every integer can be written either as $2q$ or $2q + 1$, for some integer $q$, because the only possible remainders after dividing by 2 are 0 and 1.

**Definition.** A **rational number** $r$ (also called a *fraction*) is a ratio of two integers; that is, $r = p/q$, where both $p$ and $q$ are integers and $q \neq 0$. A real number that is not rational is called **irrational**.

It turns out that there are plenty of irrational numbers, as we shall soon see. The terms *rational* and *irrational* come from "ratio;" in particular, this usage of irrational does not mean "unreasonable," which is the other contemporary usage of this word.

There are many ways to write a given rational number; for example, $\frac{1}{2} = \frac{2}{4} = \frac{3}{6} = \cdots$. Recall that

$$\frac{a}{b} = \frac{c}{d}$$

if and only if **cross multiplication** holds: $ad = bc$. Note that cross multiplying converts equations in rationals into equations in integers. Given a rational number $a/b$, let us show that we may assume, after suitable cancellation, that at least one of the numerator $a$ and the denominator $b$ is odd. Theorem 1.16 says that

$$a = 2^\ell m \quad \text{and} \quad b = 2^k n,$$

where $\ell, k \geq 0$ and $m$ and $n$ are odd. If $\ell \geq k$, then

$$\frac{a}{b} = \frac{2^\ell m}{2^k n} = \frac{2^{\ell-k} m}{n}.$$

Because $\ell \geq k$, the exponent $\ell - k \geq 0$ and $2^{\ell-k} m$ is an integer. We may thus replace $a$ by $2^{\ell-k} m$ and $b$ by $n$; that is, we may assume that the (new) denominator is odd. Similarly, if $\ell \leq k$, then

$$\frac{a}{b} = \frac{m}{2^{k-\ell} n},$$

and we may assume that the (new) numerator is odd. Hence, every rational number $a/b$ has an expression of the form $p/q$, where at least one of $p$ and

$q$ is odd. (More is true; $a/b$ can always be put in *lowest terms*; that is, the numerator and denominator have no factors in common. This is proved using the notion of *greatest common divisor*, but we leave this discussion to another course.)

***Theorem 2.5.*** $\sqrt{2}$ is an irrational number.

*Proof.* Assume, on the contrary, that $\sqrt{2}$ is rational; that is,

$$\sqrt{2} = \frac{p}{q},$$

where both $p$ and $q$ are positive integers. By our discussion above, we may assume that at least one of $p$ and $q$ is odd.

Squaring both sides, $2 = p^2/q^2$, and cross multiplying gives

$$2q^2 = p^2.$$

Were $p$ odd, then $p^2$ would also be odd [by Exercise 2.4(ii): the product of odds is odd]. Since $p^2 = 2q^2$ is even, we conclude that $p$ is even, and so we may write $p = 2r$ for some integer $r$. Substituting into $2q^2 = p^2$ gives $2q^2 = (2r)^2 = 4r^2$, so that

$$q^2 = 2r^2.$$

It follows, as above, that $q$ is even (for $q^2 = 2r^2$ is even). This contradicts our assumption that at least one of $p$ and $q$ is odd. •

This last result is significant in the history of mathematics. We accept the real number line without qualms; if two points on a line are chosen, one labeled 0 and the other 1, then every point on the line corresponds to a number. Not only was this not obvious to the ancient Greeks, they did not believe it. From its earliest days, about four to five thousand years ago in western Asia, mathematics was used for applications to practical problems. Numbers and geometry were studied using experience, and generalizations seem to have been made with the light of intuition, but not with proofs. Because of its utility, mathematics spread to China, India, Greece, and Egypt. In Greece, sometime after the time of Homer, the idea arose that logical reasoning was necessary to prove mathematical results. There are proofs attributed to Thales of Miletus (ca. 624–547 BC) (many historians describe him as the founding father of Greek mathematics). The ancient philosopher Pythagoras (ca. 570–500 BC), after whom the Pythagorean theorem is named, may have studied with Thales. Pythagoras founded a secret philosophical society, one of whose goals was to use integers and their ratios to explain all of nature's phenomena. (The word *mathematics*, meaning "that which is learned," is due to the Pythagoreans. Here is some more etymology. The word *geometry*, meaning "earth measure," probably goes back to its early applications, whereas the word *algebra* is a European version of the first word in the title of an influential book, *Al-jabr w'al muqabala*[7], by al-Khwarizmi, written in Arabic in 830.) For the Pythagoreans, numbers were defined to be positive integers, whereas other (positive) real numbers were not numbers at all; instead, they were viewed as pairs $a : b$ of line segments (which we can interpret as the number $|a|/|b|$, where $|a|$ is the length of $a$). There were geometric ways of viewing addition, subtraction, multiplication, and division of segments, but it was virtually impossible to do any algebra with them. For example, a sophisticated geometric argument (due to Eudoxus and given in Euclid's *Elements*) was needed to prove cross multiplication: if $a : b = c : d$, then $a : c = b : d$. Pythagoras dealt with rationals by assuming, given two segments $x$ and $y$, that there is a segment $z$ and integers $m$ and $n$ with $|x| = m|z|$ and $|y| = n|z|$; in modern notation: $|x| = \frac{m}{n}|y|$. He had hoped that such a relation would be true for any pair of segments $x$ and $y$, but Theorems 2.3 and 2.5 showed him that this is not so when $x$ is the diagonal of a square with side $y$.

Why did Pythagoras have such a constrictive view of numbers? We quote van der Waerden, *Science Awakening*.

> Nowadays we say that the length of the diagonal is the "irrational number" $\sqrt{2}$, and we feel superior to the poor Greeks who "did not know irrationals." But the Greeks knew irrational ratios very well. ... That they did not consider $\sqrt{2}$ as a number was not a result of ignorance, but of strict adherence to the definition of number. *Arithomos* means quantity, therefore whole number. Their logical rigor did not even allow them to admit fractions; they replaced them by ratios of integers.

> For the Babylonians, every segment and every area simply represented a number. ... When they could not determine a square root exactly, they calmly accepted an approximation. Engineers and natural scientists have always done this. But the Greeks were concerned with exact knowledge, with "the diagonal itself," as Plato expresses it, not with an acceptable approximation.

---

[7]One can translate this title from Arabic, but the words already had a technical meaning: both *jabr* and *muqabala* refer to certain operations akin to subtracting the same number from both sides of an equation.

In the domain of numbers (positive integers), the equation $x^2 = 2$ cannot be solved, not even in that of ratios of numbers. But it is solvable in the domain of segments; indeed the diagonal of the unit square is a solution. Consequently, in order to obtain exact solutions of quadratic equations, we have to pass from the domain of numbers (positive integers) to that of geometric magnitudes. Geometric algebra is valid also for irrational segments and is nevertheless an exact science. It is therefore logical necessity, not the mere delight in the visible, which compelled the Pythagoreans to transmute their algebra into a geometric form.

Even though the Pythagorean definition of number is no longer popular, the Pythagorean dichotomy persists to this very day. For example, almost all American high schools teach one year of algebra and one year of geometry, instead of two years in which both subjects are developed together. The problem of defining *number* has arisen several times since the classical Greek era. In the 1500's, mathematicians had to deal with negative numbers and with complex numbers (see our discussion of cubic polynomials in Chapter 4); the description of real numbers generally accepted today dates from the late 1800's. There are echos of Pythagoras in our time. L. Kronecker (1823–1891) wrote, "Die ganzen Zahlen hat der liebe Gott gemacht; alles andere ist Menschenwerk" (The integers were created by God; all the rest is the work of Man), and even today, some logicians argue for a new definition of number.

### Exercises

**2.1.** (i) In Figure 2.11*a*, $QPSR$ is a rectangle and $YZSR$ is a parallelogram. Show that $\triangle QYR$ and $\triangle PZS$ are congruent.

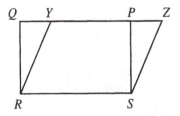

Figure 2.11*a*

(ii) Show that one can construct the parallelogram $YZSR$ from the rectangle $QPSR$ by cutting off $\triangle QYR$ and pasting it in position $\triangle PZS$; conclude that the parallelogram has the same area as the rectangle if $Y$ is between $Q$ and $P$.
(iii) Prove that the parallelogram $YZSR$ has the same area as the rectangle $QPSR$ when $Y$ is not between $Q$ and $P$ (see Figure 2.11$b$.) [One must prove this in order to complete the argument that the areas of the parallelogram and the rectangle always agree. You may not use Theorem 2.2, for this exercise is to give an alternative proof of that theorem.]

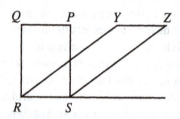

Figure 2.11$b$

**2.2.** Show that the trapezoid in Figure 2.12 has area $\frac{1}{2}(a+b)h$.

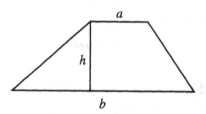

Figure 2.12

**2.3.** Assume, in Figure 2.13, that $PA$, $EH$, and $QR$ are parallel. If $P$ and $Q$ are points, let us denote the length of the line segment $PQ$ by $|PQ|$. Prove that $|EF| = |GH|$, and then conclude that Cavalieri's Principle applies to $\triangle PQR$ and $\triangle AQR$. (Hint: Let $\ell$ and $\ell'$ be parallel lines, and let $t$ and $t'$ be transversals. If $\ell''$ is parallel to $\ell$ (and to $\ell'$), then $\ell''$ divides the transversals proportionally. In Figure 2.13, $|PE|/|PQ| = |AH|/|AR|$.)

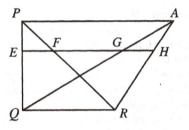

Figure 2.13

**2.4.** Let $a$ and $b$ be integers.
   (i) If $a$ is even, prove that $ab$ is even for every integer $b$.
   (ii) If both $a$ and $b$ are odd, prove that $ab$ is odd while $a + b$ is even.
   (iii) If neither $a$ nor $b$ is a multiple of 3, prove that $ab$ is not a multiple of 3.

**2.5.** If $r = p/q$ is a nonzero rational number, show that $r + \sqrt{2}$ and $r\sqrt{2}$ are irrational numbers. Conclude that there are infinitely many irrational numbers.

**2.6.** Use the Pythagorean theorem to prove that if $a$ is the side length of a cube and $|AB|$ is the length of a diagonal joining opposite corners, then $|AB|^2 = 3a^2$.

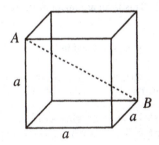

Figure 2.14

**2.7.** Prove that $\sqrt{3}$ is irrational. (Hint: Modify the proof of Theorem 2.5, replacing each occurrence of 2 by 3, "even" by "multiple of 3", and "odd" by "not divisible by 3.")

*Remark*: Exercise 2.7 shows one reason why it is important to know proofs. One cannot use the statement of Theorem 2.5 to solve the problem. However, one can solve this problem by modifying the proof of that theorem.

**2.8.** (i) Prove that an integer $m \geq 2$ is a perfect square if and only if each of its prime factors occurs an even number of times. (Hint: Use the *Fundamental*

**Theorem of Arithmetic**: If $p_1^{e_1} \cdots p_n^{e_n} = q_1^{f_1} \cdots q_t^{f_t}$, where $p_1 < p_2 < \cdots < p_n$ and $q_1 < q_2 < \cdots < q_t$ are primes and the $e$'s and $f$'s are positive integers, then $n = t$ and, for all $i$, $p_i = q_i$ and $e_i = f_i$. [This theorem is usually proved in the next course.])

(ii) Using part (i), prove that if $m$ is a positive integer for which $\sqrt{m}$ is rational, then $m$ is a perfect square. Conclude that $\sqrt{2}, \sqrt{3}$, and $\sqrt{6}$ are irrational.

(iii) If $n$ is a positive integer, show that $n^3 + n^2$ is a perfect square if and only if $n + 1$ is a perfect square. (Example: If $n = 8$ (so that $n + 1 = 9 = 3^2$), then $n^3 + n^2 = 512 + 56 = 576 = 24^2$. The next example occurs when $n = 15$.)

**2.9.** Let $p$ be a prime number, and consider the number

$$N = N_p = 1 + (2 \times 3 \times 5 \times 7 \times 11 \times \cdots \times p).$$

Prove that none of the prime numbers $2, 3, 5, 7, 11, \ldots, p$ used in the definition of $N$ is a divisor of $N$. (Hint: Dividing $N$ by 3, for example, leaves remainder 1, for the quotient is $2 \times 5 \times 7 \times 11 \times \cdots \times p$.)

**2.10.** Use Exercise 2.9 to prove that there are infinitely many prime numbers. (Hint: Assume, on the contrary, that there are only a finite number of primes, say, $2, 3, 5, 7, \ldots, p$; define $N$ as in Exercise 2.9 and show that $N$ does not satisfy Theorem 1.2.) (This is an argument given in the ninth book of Euclid's *Elements*.)

**2.11.** (i) If $p = 11$, the number $N$ defined in Exercise 2.9 is 2311. Show that 2311 is prime.

(ii) If $p = 13$, the number $N$ is 30031. Show that 30031 is not prime.

(iii) If $p = 17$, show that 19 is a divisor of $N = 510511$.

**2.12.** A mad architect has designed the symmetric building shown in Figure 2.15. Find the area of the building's front (not counting the two circular windows of radius 2 or the semicircular entrance way), given the dimensions in the figure.

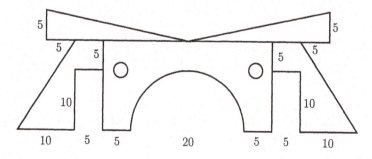

Figure 2.15

## THE PYTHAGOREAN THEOREM

Euclid alone has looked on beauty bare.
*Edna St. Vincent Millay*

From earliest times, algebraic identities were verified by geometric figures. The simplest example is Figure 2.16, which is a geometric picture of the identity $(a + b)^2 = a^2 + 2ab + b^2$;

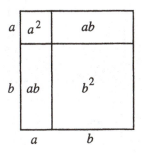

Figure 2.16

the large square having sides of length $a + b$ is divided into two squares, with side lengths $a$ and $b$, respectively, and two rectangles each of area $ab$.

Recall that the *hypotenuse* (from the Greek word meaning "to stretch") of a right triangle is the longest of its three sides; the other sides are called its *legs*. In Theorem 2.3, we proved the special case of the Pythagorean theorem involving an isosceles right triangle.

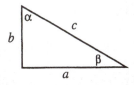

Figure 2.17

***Theorem 2.6 (Pythagorean Theorem).*** In a right triangle with legs of lengths $a$ and $b$ and hypotenuse of length $c$, we have $c^2 = a^2 + b^2$.

*Remark.* Although the statement of this theorem was accepted centuries before him, Pythagoras was perhaps the first to prove it. The proof we give, due to Indian mathematicians around 400 AD, is based on Figure 2.18.

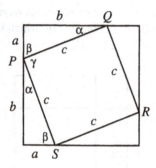

Figure 2.18

*Proof.* Figure 2.18 pictures the area of the big square in two ways: first, as a square with side lengths $a + b$; second, as dissected into a rhombus $PQRS$ with side lengths $c$, and four congruent right triangles of area $\frac{1}{2}ab$. We claim that the rhombus is actually a square. Consider the interior angle $\gamma$ at $P$, for example. Note that $\alpha + \gamma + \beta = 180°$. Inasmuch as $\alpha + \beta = 90°$, because $\alpha$ and $\beta$ are the acute angles in a right triangle, we have $\gamma = 90°$. The Pythagorean theorem now follows from the algebraic identity

$$(a + b)^2 = c^2 + 4 \times \tfrac{1}{2}ab,$$

for the left side is $a^2 + 2ab + b^2$, while the right side is $c^2 + 2ab$.   •

The converse of the Pythagorean theorem is also true.

***Theorem 2.7.*** A triangle having sides of lengths $a$, $b$, and $c$ with $a^2 + b^2 = c^2$ must be a right triangle.

*Proof.* Take two perpendicular lines, as in Figure 2.19, and choose points $A$ and $B$ with $|AO| = a$ and $|BO| = b$. Now $\triangle OAB$ is a right triangle having

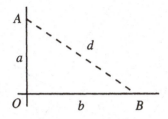

Figure 2.19

side lengths $a$, $b$, and $d$. The Pythagorean theorem gives $a^2 + b^2 = d^2$, and so $d = c$. But $\triangle OAB$ and the given triangle are congruent, by "side-side-side," and so $\triangle$ is a right triangle.  •

Figure 2.20 gives another proof, also in the Indian style, of the Pythagorean theorem.

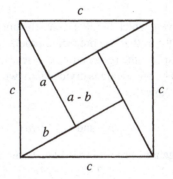

Figure 2.20

The square of side length $c$ is partitioned, with a square of side length $a - b$ in the center. This yields

$$c^2 = (a - b)^2 + 4 \times \tfrac{1}{2}ab$$

for the total area. The Pythagorean theorem follows. Query: What does Figure 2.20 look like when the right triangle is isosceles?

The proof of the Pythagorean theorem given in Euclid's *Elements* is based on Figure 2.21. Note that it is almost purely geometric (in contrast to

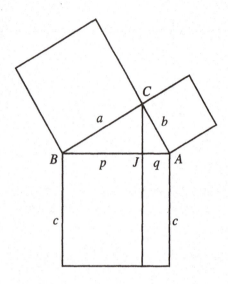

Figure 2.21

the two proofs just given, which involve some algebra as well), and, thus, it is more complicated for us than the preceding proofs. On the other hand, notice that it actually displays the (geometric) squares sitting on the three sides of the right triangle. The big square is divided into two rectangles of areas $pc$ and $qc$. It suffices to show that

$$a^2 = pc \quad \text{and} \quad b^2 = qc.$$

There is an algebraic proof of these equations. The altitude $CJ$ forms similar right triangles $\triangle JBC$, $\triangle CBA$, and $\triangle JCA$, as in Figure 2.22. The

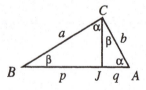

Figure 2.22

corresponding sides of these similar triangles can be seen in Figure 2.23. Thus, there are the proportions

$$\frac{p}{a} = \frac{a}{c} \quad \text{and} \quad \frac{q}{b} = \frac{b}{c},$$

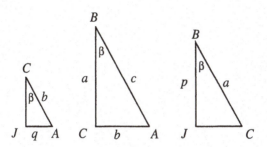

Figure 2.23

and cross multiplication gives the desired equations.

At this stage, however, Euclid did not want to use proportions. Although such algebraic manipulations are routine for us, $p : a = a : c$ in Euclid's time was not merely an equation $p/a = a/c$ (as we remarked earlier); rather, it was a relation between two pairs of line segments. Geometry and algebra were already living in separate worlds, and Euclid's treatment of proportions is very sophisticated. Thus, Euclid was not being a purist in avoiding proportions; his proof, which looks more complicated to us, was the most straightforward one that he knew.

Euclid's geometric argument is best explained by referring to Figure 2.24.

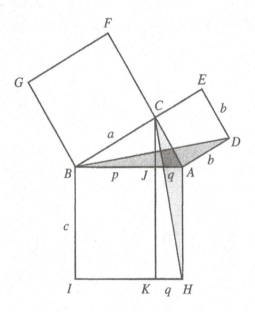

Figure 2.24

The area $c^2$ of the square $\square AHIB$ is the sum of the areas of the rectangles $\square AHKJ$ and $\square BJKI$. We want to show that

$$\text{area}(\square AHKJ) = b^2 \quad \text{and} \quad \text{area}(\square BJKI) = a^2,$$

which will prove the desired result.

Now, by Theorem 2.1,

$$\text{area}(\triangle CAH) = \tfrac{1}{2}|AH||AJ|,$$

for $AH$ is a base and $AJ$ is the corresponding altitude. Thus,

$$\text{area}(\triangle CAH) = \tfrac{1}{2}\,\text{area}(\square AHKJ).$$

Next we compare $\triangle CAH$ with $\triangle ABD$. The angles $\angle CAH$ and $\angle DAB$ are equal, for each equals $\angle BAC + 90°$. By the side-angle-side theorem, it follows that $\triangle CAH$ and $\triangle ABD$ are congruent. But

$$\text{area}(\triangle ABD) = \tfrac{1}{2}|AD||AC| = \tfrac{1}{2}b^2,$$

for $AD$ is the base and $AC$ is the altitude. It follows that

$$\text{area}(\square AHKJ) = 2\,\text{area}(\triangle CAH) = 2\,\text{area}(\triangle ABD) = b^2.$$

A similar argument shows that

$$\text{area}(\square BJKI) = a^2,$$

and the Pythagorean theorem follows.   •

We now give a geometric proof of the Pythagorean theorem that Euclid would have loved had he known it[8] (this proof can be found in a commentary in Heath's 1926 translation of Euclid). Figure 2.25 displays a square of side length $a+b$ dissected in two different ways. The left dissection is Figure 2.18; the right dissection is Figure 2.16 with each rectangle bisected by a diagonal.

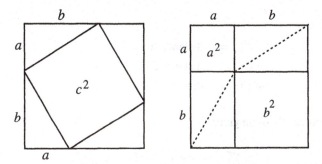

Figure 2.25

That both squares have the same area gives the equation

$$c^2 + 4 \times \tfrac{1}{2}ab = a^2 + b^2 + 4 \times \tfrac{1}{2}ab.$$

Here are some applications of the Pythagorean theorem. In a later section of this chapter, we will discuss a bit of trigonometry (which is also intimately related to the Pythagorean theorem).

---

[8]There is a book by Elisha Scott Loomis, "The Pythagorean Proposition," Edwards Bros, Ann Arbor, Michigan, 1940, containing 370 [by the author's count] different proofs of the Pythagorean theorem!

Aristarchus (310–250 BC) used the Pythagorean theorem to draw conclusions about the distance from the Moon to the Earth. His idea was as follows. At halfmoon, the Sun $S$, the Moon $M$, and the Earth $E$ form a right triangle, with right angle at $M$, as in Figure 2.26.

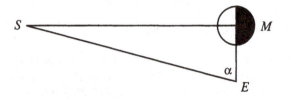

Figure 2.26

Let

$$
\begin{aligned}
|SE| &= \text{distance from } S \text{ to } E, \\
|SM| &= \text{distance from } S \text{ to } M, \\
|ME| &= \text{distance from } M \text{ to } E.
\end{aligned}
$$

The Pythagorean theorem gives

$$|SE|^2 = |SM|^2 + |ME|^2.$$

One conclusion Aristarchus drew from this is that the Earth is farther from the Sun than it is from the Moon. This is not at all obvious, for both the Sun and the Moon appear to be the same size (of course, having observed solar eclipses in which the Moon is seen to come between the Earth and the Sun, the Greeks would have known this fact without using the Pythagorean theorem). At sunset, if one is looking up at the (half) Moon, sunlight seems to be perpendicular to $EM$, the line of sight to the Moon; that is, the angle $\alpha$ seems very close to 90°. Now $\cos \alpha = |ME|/|SE|$. When $\alpha$ is close to 90°, $\cos \alpha$ is close to zero. Aristarchus was, thus, able to conclude that the Earth is very much farther from the Sun than it is from the Moon, and so this fact follows without sophisticated instruments. [Modern measurements have $|ME| \approx 240,000$ miles and $|SE| \approx 93,000,000$ miles, so that $\cos \alpha = |ME|/|SE| \approx .0026$ and $\alpha \approx 89.85°$ (or $89°51'$).]

The following *castle problem* is from an old Chinese text.

There is a circular castle, whose diameter is unknown; it is provided with four gates and two lengths out of the north gate there is a large tree, which is visible from a point six lengths east of the south gate. What is the length of the diameter?

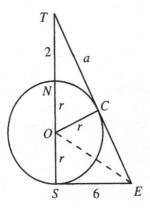

Figure 2.27

We compute the area of $\triangle TSE$ in two ways. On the one hand,

$$\text{area}(\triangle TSE) = \tfrac{1}{2}(2 + 2r)6 = 6 + 6r.$$

On the other hand, this area is the sum of the areas of the three smaller triangles $\triangle OSE$, $\triangle OCE$, and $\triangle OCT$. Now the right triangles $\triangle OSE$ and $\triangle OCE$, being congruent, have the same area, namely, $\tfrac{1}{2}6r = 3r$, whereas $\text{area}(\triangle OCT) = \tfrac{1}{2}ra$, where $a = |CT|$. We conclude that

$$6 + 6r = 3r + 3r + \tfrac{1}{2}ra;$$

that is,

$$12 = ra.$$

The Pythagorean theorem (applied to $\triangle OCT$) gives

$$r^2 + a^2 = (r + 2)^2$$

$$= r^2 + 4r + 4,$$

and so

$$a^2 = 4r + 4.$$

Since $12 = ra$,

$$\frac{144}{r^2} = 4r + 4,$$

which simplifies to the cubic

$$r^2(r + 1) = 36.$$

We can solve this equation by trial and error, evaluating the left-hand side for small values of $r$; if $r = 1$, then $r^2(r+1) = 2$; if $r = 2$, then $r^2(r+1) = 12$; if $r = 3$, then $r^2(r+1) = 36$. Thus, the radius of the castle is $r = 3$ lengths, and so the diameter of the castle is 6 lengths. This problem was solved by the Chinese mathematician Ch'in Chiu-shao in 1247. We shall examine cubic polynomials in Chapter 4.

*Exercises*

**2.13.** Use a dissection of a cube having side lengths $a + b$ to prove

$$(a + b)^3 = a^3 + 3a^2b + 3ab^2 + b^3.$$

**2.14.** Give another proof of the Pythagorean theorem (attributed to the U.S. President James A. Garfield). Take a vertical line $CC'$ of length $a+b$, and construct two replicas of $\triangle ABC$ as in Figure 2.28. (Hint: Check that $\angle BAB'$ is a right angle.)

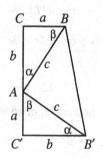

Figure 2.28

**2.15.** (i) In a triangle with sides of lengths 13, 14, and 15, what is the length of the altitude drawn to the side of length 14?
(ii) Find the area of this triangle.

**2.16.** Given a right triangle with perpendicular sides $a$ and $b$, find the side $s$ of the inscribed square.

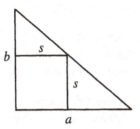

Figure 2.29

**2.17.** Given a right triangle with legs $a$ and $b$, find the radius $r$ of the inscribed circle.

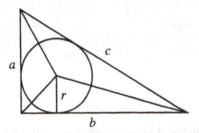

Figure 2.30

**2.18.** The length of the perimeter of a right triangle is 60 units, and the length of the altitude perpendicular to the hypotenuse is 12 units. How long are the sides?

**2.19.** A cylindrical column has height 30 feet and circumference 8 feet. A garland is wound evenly 5 times around the column, reaching from top to bottom. Show that the garland must be at least 50 feet long.

**2.20.** (i) Show that an equilateral triangle $\triangle$ with side lengths $a$ has altitudes of height $a\sqrt{3}/2$.
(ii) Show that the area of $\triangle$ is $a^2\sqrt{3}/4$.

**2.21.** Find the area and circumference of the circumscribed circle of a regular hexagon whose sides have length $s$.

**2.22.** Find the area and circumference of the circumscribed circle of an equilateral triangle whose sides have length $s$. (Hint: The center of the circumscribed circle lies on an altitude of the triangle $\frac{2}{3}$ the length from the vertex.)

## PYTHAGOREAN TRIPLES

"When I use a word," Humpty Dumpty said, in a rather scornful tone, "it means just what I choose it to mean—neither more nor less."

*Lewis Carroll*

What are the right triangles all of whose sides have integer length? Almost everyone is familiar with the fact that $3^2 + 4^2 = 5^2$, and many know that $5^2 + 12^2 = 13^2$. Are there other positive integers $a$, $b$, $c$ with $a^2 + b^2 = c^2$?

*Definition.* A *Pythagorean triple* is a triple $(a, b, c)$ of positive integers $a < b < c$ with $a^2 + b^2 = c^2$. The smaller numbers $a$ and $b$ in a Pythagorean triple are called its *legs*.

It is no loss in generality to assume that the two legs $a$ and $b$ in a Pythagorean triple $(a, b, c)$ are always distinct, for if $a = b$, then $c^2 = a^2 + a^2 = 2a^2$, hence $2 = (c/a)^2$, and $\sqrt{2} = c/a$, contradicting the irrationality of $\sqrt{2}$.

On a Babylonian cuneiform tablet dating back to about 1800 BC (nowadays called Plimpton 322), one finds a list of Pythagorean triples, one of which is

$$(12\,709,\ 13\,500,\ 18\,541).$$

There is also archaeological evidence of Pythagorean triples being used in China about 200 BC and in India about 500 BC. Historians today believe that the Babylonians were aware of a method for generating Pythagorean triples, although it is not clear that they knew whether the method generated all of them. Indeed, here is a simple way to find many Pythagorean triples. Begin by writing

$$a^2 = c^2 - b^2 = (c + b)(c - b).$$

This equation can be solved if both factors $c + b$ and $c - b$ are squares; say, there are integers $m$ and $n$ with

$$c + b = m^2$$

and

$$c - b = n^2$$

Given $m$ and $n$, we can now find $b$ and $c$.

$$c = \tfrac{1}{2}(m^2 + n^2)$$

$$b = \tfrac{1}{2}(m^2 - n^2);$$

if $m$ and $n$ are both odd, so are their squares; the sum and difference of odds are even, so that both $b$ and $c$ are integers, as required. Finally,

$$a = \sqrt{c^2 - b^2} = mn.$$

With these formulas, every choice of odd numbers $m > n$ gives a Pythagorean triple. It is clear that one can produce "big" Pythagorean triples in this way; for example, $m = 179$ and $n = 71$ give the big triple on Plimpton 322. This method does not give all possible Pythagorean triples, however. For example, $(9, 12, 15)$ does not arise in this way: the only candidates for odd numbers $m > n$ with $9 = mn$ are $m = 9$ and $n = 1$, but $\tfrac{1}{2}(9^2 - 1^2) = 40$ and $\tfrac{1}{2}(9^2 + 1^2) = 41$.

Right triangles were of interest in ancient civilizations for several reasons: measurement (trigonometry, based on right triangles, was very useful to them), astronomy, and religion (astronomy is intimately involved with calendars, and dates are vital for predicting the seasons; this was important both for agricultural and religious reasons). Triangles with integral sides are, clearly, the most convenient ones for calculations (decimal notation is a recent invention), and it seems most likely that the ancients emphasized them because of their importance in teaching these subjects.[9] For example, here is a problem from *Nine Chapters on the Mathematical Art* written in China about 2000 years ago. (Similar problems can be found in ancient manuscripts from other cultures as well.)

> There is a door whose height and width are unknown, and a pole whose length ($= p$) is also unknown. Carried horizontally, the pole does not fit by 4 ch'ih; vertically, it does not fit by 2 ch'ih; slantwise, it fits exactly. What are the height, width, and diagonal of the door?

---

[9]It is obvious, however, that the writer of Plimpton 322 was not interested only in utilitarian matters.

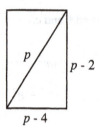

Figure 2.31

The Pythagorean theorem gives the equation $(p-4)^2+(p-2)^2 = p^2$, which simplifies to $p^2 - 12p + 20 = (p-10)(p-2) = 0$. It follows that the pole is $p = 10$ ch'ih long and that the dimensions of the door are 8 ch'ih and 6 ch'ih (the other root $p = 2$ does not fit the physical data). If one is interested in teaching applications of the Pythagorean theorem, it would be foolish to pose this problem so that the pole has length $p = 12$ ch'ih, for there is no Pythagorean triple with $c = 12$, as we shall see in a moment. A student trying to solve this variant of the problem would have to compute an irrational square root, which obviously complicates the problem.

We shall see, in Exercise 2.32, that every integer $m \geq 3$ occurs as a leg of some Pythagorean triple. On the other hand, not every positive integer $c$ occurs in some Pythagorean triple $(a, b, c)$, for $c$ need not be a sum of two squares. For example, can $c = 12$? Are there positive integers $a < b$ with

$$a^2 + b^2 = 144?$$

After Theorem 1.6, we showed that there are $\frac{1}{2}(10 \times 11) = 55$ possible sums $a^2 + b^2$ to check, but a preliminary discussion will reduce the number of cases considerably. Because $b \leq 11$, the possible values for $b^2$ are

$$1, 4, 9, 16, 25, 36, 49, 64, 81, 100, 121.$$

As $a < b$, we have $a^2 < b^2$, so that

$$144 = a^2 + b^2 < 2b^2.$$

Hence,

$$72 < b^2,$$

and the only possible values for $b^2$ are 81, 100, and 121. If $b^2 = 121$, then

$$144 - 121 = 23;$$

since 23 is not a perfect square, $b^2 \neq 121$ and $b \neq 11$. Similarly, if

$$b^2 = 81 \quad \text{or} \quad b^2 = 100,$$

then $144 - 81 = 63$ or $144 - 100 = 44$, neither of which is a perfect square. We have proved the geometric result that there is no right triangle with hypotenuse of length 12 both of whose legs have integral length.

Let us try to repeat this argument for $c = 13$. Are there positive integers $a < b$ with

$$a^2 + b^2 = 13^2 = 169?$$

As above, $169 = a^2 + b^2 < 2b^2$, and hence,

$$84.5 = 169/2 < b^2.$$

The only possibilities are

$$b^2 = 100, 121, 144.$$

Since

$$169 - 100 = 69$$

$$169 - 121 = 48$$

$$169 - 144 = 25 = 5^2,$$

we conclude that $b = 12$; this gives the Pythagorean triple $(5, 12, 13)$; moreover, we see that this is the only Pythagorean triple with $c = 13$. In seeking other Pythagorean triples, it is unlikely that we would get very far using only these primitive techniques.

A man bored with the Pythagorean triple $(3, 4, 5)$ would be happy to discover $(5, 12, 13)$, but he would not really be satisfied with $(6, 8, 10)$; the latter arises from $(3, 4, 5)$ merely by doubling each of the numbers. In the same way, every Pythagorean triple $(a, b, c)$ gives rise to infinitely many "new" Pythagorean triples: if $\ell$ is a positive integer, then $(\ell a, \ell b, \ell c)$ is also a Pythagorean triple, for

$$(\ell a)^2 + (\ell b)^2 = \ell^2 a^2 + \ell^2 b^2 = \ell^2 (a^2 + b^2) = \ell^2 c^2 = (\ell c)^2.$$

Thus, $(3, 4, 5)$ generates infinitely many Pythagorean triples $(3\ell, 4\ell, 5\ell)$, but they are merely variants of it.

Let us look a bit further. Both (6, 8, 10) and (9, 12, 15) are Pythagorean triples arising from (3, 4, 5), but neither arises from the other by multiplying all three numbers by an integer $\ell$. However, there are positive integers $\ell(= 3)$ and $m(= 2)$ with $(6\ell, 8\ell, 10\ell) = (9m, 12m, 15m)$.

The somewhat vague question we are considering is when two Pythagorean triples are "genuinely" different. We have been led to the following definition.

**Definition.** Two Pythagorean triples $(a, b, c)$ and $(a', b', c')$ are **similar** if there are positive integers $\ell$ and $m$ with

$$\ell a = ma', \quad \ell b = mb', \quad \text{and} \quad \ell c = mc'.$$

It is more natural to look at this definition geometrically; after all, every Pythagorean triple $(a, b, c)$ does arise as the side lengths of a right triangle, by the converse of the Pythagorean theorem. Let Pythagorean triples $(a, b, c)$ and $(a', b', c')$ arise from right triangles $\triangle$ and $\triangle'$, respectively. If the triples are similar, then $\triangle$ and $\triangle'$ are similar triangles, for their sides are proportional (the constant of proportionality is $m/\ell$). Conversely, if the triangles are similar, then corresponding sides are proportional; that is, there is a number $r$ with $a/a' = r = b/b' = c/c'$. As all the side lengths now are integers, it follows that $r$ is a rational number: $r = m/\ell$, where $m$ and $\ell$ are positive integers. Therefore, $(a, b, c)$ and $(a', b', c')$ are similar Pythagorean triples.

Intuitively, we regard similar Pythagorean triples as being "essentially the same" and non-similar Pythagorean triples as being "genuinely different."

Here are some more examples of Pythagorean triples; Exercise 2.25 asks you to show that no two of these are similar.

$$
\left.
\begin{aligned}
&(3, 4, 5); \\
&(5, 12, 13); \\
&(7, 24, 25); \\
&(8, 15, 17); \\
&(781, 2460, 2581) \\
&(3993, 7972024, 7972025).
\end{aligned}
\right\} \quad (1)
$$

None of the examples of Pythagorean triples we have seen have two odd legs; are there any such?

***Theorem 2.8.***(i) There is no Pythagorean triple having both legs odd.
(ii) Every Pythagorean triple $(a, b, c)$ is similar to one which has one leg odd and the other leg even.

*Proof.* (i) Suppose there is a Pythagorean triple $(a, b, c)$ with $a = 2m + 1$ and $b = 2n+1$. Then both $a^2$ and $b^2$ are odd, so that $c^2$ is even, by Exercise 2.4(ii). It now follows from this exercise that $c$ is even; say, $c = 2r$. Therefore, we can rewrite the equation $a^2 + b^2 = c^2$ as

$$(2m + 1)^2 + (2n + 1)^2 = (2r)^2.$$

After expanding, the left side is $4(m^2 + m + n^2 + n) + 2$, and the right side is $4r^2$. After dividing by 2, the left side is $2(m^2+m+n^2+n)+1$, an odd number, whereas the right side is $2r^2$, an even number, and this is a contradiction.
(ii) We have just eliminated the possibility that both $a$ and $b$ are odd. If both $a$ and $b$ are even, say, $a = 2s$ and $b = 2t$, then $c$ must also be even, say, $c = 2u$. The equation $a^2 + b^2 = c^2$ gives

$$4s^2 + 4t^2 = 4u^2.$$

Canceling 4's shows that $(s, t, u)$ is also a Pythagorean triple, which is clearly similar to $(a, b, c)$. If one of $s$ and $t$ is odd, we are done [for the other must be even, by part (i)]. If both $s$ and $t$ are even, we may repeat what we have just done. By Theorem 1.16, $s = 2^k m$, where $k \geq 0$ and $m$ is odd. Thus, one needs at most $k$ repetitions before arriving at an odd leg.  •

In the next section, we will derive a formula giving all Pythagorean triples $(a, b, c)$ which are not similar: choose positive integers $q > p$, let $q^2 - p^2$ and $2qp$ be the legs, and let $c = q^2 + p^2$. It is easy to check that

$$(q^2 - p^2)^2 + (2qp)^2 = (q^2 + p^2)^2,$$

so that either $(q^2 - p^2, 2qp, q^2 + p^2)$ or $(2qp, q^2 - p^2, q^2 + p^2)$ is a Pythagorean triple [depending on whether $q^2 - p^2 < 2qp$ or $2qp < q^2 - p^2$]. We will also see that every Pythagorean triple is similar to such a Pythagorean triple. For example, if $q = 2$ and $p = 1$, then

$$2^2 - 1^2 = 3, \quad 2 \times 2 \times 1 = 4, \quad 2^2 + 1^2 = 5,$$

and we have the Pythagorean triple $(3, 4, 5)$.

If $q = 3$ and $p = 2$, then

$$3^2 - 2^2 = 5, \quad 2 \times 3 \times 2 = 12, \quad 3^2 + 2^2 = 13,$$

and we have the Pythagorean triple (5, 12, 13).

If $q = 3$ and $p = 1$, then

$$3^2 - 1^2 = 8, \quad 2 \times 3 \times 1 = 6, \quad 3^2 + 1^2 = 10,$$

and we have the Pythagorean triple (6, 8, 10).

If $q = 4$ and $p = 1$, then

$$4^2 - 1^2 = 15, \quad 2 \times 4 \times 1 = 8, \quad 4^2 + 1^2 = 17,$$

and we have the Pythagorean triple (8, 15, 17).

Notice that $q^2 - p^2 < 2qp$ in the first two examples, while $2qp < q^2 - p^2$ in the last two examples. We shall discuss the reason for this in the next section.

The $q > p$ formula can be used to check whether an alleged Pythagorean triple really is one. Let us show (again) that (12709, 13500, 18541), the Babylonian triple mentioned earlier, is a Pythagorean triple. Of course, we may check that $12709^2 + 13500^2 = 18541^2$, but the numbers are rather large. An alternative method is to find $q$ and $p$. Because $2qp$ is even, $2qp = 13500$ and so the odd leg is $q^2 - p^2 = 12709$. As $q^2 + p^2 = 18541$, we have

$$q^2 + p^2 = 18541$$

$$q^2 - p^2 = 12709.$$

Subtracting, $2p^2 = 5832$, $p^2 = 2916$, and $p = 54$. Since $2qp = 13500$, we have $q = 125$. Thus, $12709^2 + 13500^2 = 18541^2$ because

$$12709 = 125^2 - 54^2,$$

$$13500 = 2 \cdot 125 \cdot 54,$$

and

$$18541 = 125^2 + 54^2.$$

This method is useful when checking whether a triple of large integers forms a Pythagorean triple. For example, Exercise 2.24 asks if the last two triples in Eq. (1) are Pythagorean triples.

## *Exercises*

**2.23.** Is there a Pythagorean triple $(a, b, c)$ with $a = 1$ or 2?

**2.24.** (i) Verify that $(781, 2460, 2581)$ is a Pythagorean triple.
(ii) Verify that $(3993, 7972024, 7972025)$ is a Pythagorean triple.
(iii) Is $(169568, 1139826, 1152370)$ a Pythagorean triple?

**2.25.** Show that no two of the Pythagorean triples in Eq. (1) are similar. (Hint: If $(a, b, c)$ and $(a', b', c')$ are similar, then $c/c' = a/a' = b/b'$.)

**2.26.** Prove that if $b = 4\binom{m+1}{2}$ for some $m \geq 1$ (e.g., $b = 4, 12, 24, 40, 60, 84, \ldots$), then there is a Pythagorean triple $(a, b, c)$ with $c = b + 1$.
Conversely, prove that if there is a Pythagorean triple $(a, b, c)$ with $c = b + 1$, then there is an integer $m \geq 2$ with $b = 4\binom{m+1}{2}$.

**2.27.** Let $\Delta$ be a right triangle with legs $a$ and $b$ and hypotenuse $c$. In Exercise 2.17, one found that the radius $r$ of the inscribed circle is given by

$$r = ab/(a + b + c)$$

[$r$ is called the *inradius* of $\Delta$].
If $(a, b, c)$ is a Pythagorean triple, prove that the inradius of the corresponding right triangle is an integer.

**2.28.** If $\Delta$ is a right triangle all of whose sides have integral length, prove that the height of the altitude to the hypotenuse is a rational number.

## THE METHOD OF DIOPHANTUS

"Talking of axes," said the Duchess, "chop off her head!"

*Lewis Carroll*

We are now going to show, using a geometric method of Diophantus[10] that the formula in the last section gives essentially all Pythagorean triples. [The extant work of Diophantus shows systematic algebraic procedures and notation, but his leaps of intuition strongly suggest that he was thinking geometrically. Fermat was aware of this, and Newton explicitly described the method of Diophantus as using chords and tangents.] Thus, geometry (the Pythagorean theorem) and applied problems suggested an algebraic problem (find all Pythagorean triples), and now we shall return to geometry to solve it. Here is more evidence that the distinction between algebra and geometry is an artificial one; both are parts of the same subject.

Let $(a, b, c)$ be a Pythagorean triple. Dividing both sides of the defining equation

$$a^2 + b^2 = c^2$$

by $c^2$ gives

$$(a/c)^2 + (b/c)^2 = 1,$$

so that the triple gives an ordered pair of positive *rational numbers* $(x, y) = (a/c, b/c)$ with $x^2 + y^2 = 1$; moreover, $x = a/c < b/c = y$.

Conversely, let us see that an ordered pair $(x, y)$ of positive rational numbers with $x < y$ and $x^2 + y^2 = 1$ gives rise to a Pythagorean triple. If we write $x$ and $y$ with the same denominator, say, $x = a/c$ and $y = b/c$, where $a$, $b$, and $c$ are positive integers, then

$$1 = x^2 + y^2 = (a/c)^2 + (b/c)^2 = (a^2 + b^2)/c^2$$

and so

$$a^2 + b^2 = c^2.$$

---

[10]We know very little about the life of Diophantus; indeed, we know only that he was a mathematician who lived in Alexandria sometime between 200BC and 350 AD (most historians believe he was born after 100AD).

Inasmuch as $a < b$ (because $a/c = x < y = b/c$), we see that $(a, b, c)$ is a Pythagorean triple.

Notice, in the passage from $(a, b, c)$ to $(a/c, b/c)$, that similar Pythagorean triples yield the same ordered pair. If $(a, b, c)$ and $(a', b', c')$ are similar, then there are positive integers $\ell$ and $m$ with $\ell a = ma'$, $\ell b = mb'$, and $\ell c = mc'$; that is, $a' = \ell a/m$, $b' = \ell b/m$, and $c' = \ell c/m$. Therefore, $a'/c' = (\ell a/m)/(\ell c/m) = a/c$ and $b'/c' = (\ell b/m)/(\ell c/m) = b/c$.

The graph of the equation $x^2 + y^2 = 1$ is the **unit circle**; that is, the circle of radius 1 with center at the origin. Let us call a point $(x, y)$ in the plane a **rational point** if both its coordinates $x$ and $y$ are rational numbers. We have just shown that the problem of finding all "genuinely different" Pythagorean triples is essentially the same as that of finding all rational points $(x, y)$ on the unit circle lying in the first quadrant (after all, both $x$ and $y$ are positive) and such that $x < y$. More precisely, every Pythagorean triple is similar to one arising from a rational point as just described.

Here is the geometric method of Diophantus (of course, we write everything in modern notation using analytic geometry that was not available to him). The point $(-1, 0)$ lies on the unit circle and it is outside the first quadrant. Let us consider lines connecting $(-1, 0)$ to other points $(g, h)$ on the unit circle.

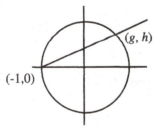

Figure 2.32

The equation of the line $\ell$ through the two points $(-1, 0)$ and $(g, h)$ is

$$y = t(x + 1),$$

where

$$t = \frac{h}{g + 1} \tag{2}$$

is the slope of $\ell$.

**Lemma 2.9.** Let $(g, h)$ be a point other than $(-1, 0)$ lying on the unit circle, and let $\ell$ be the line joining these two points. Then $\ell$ has the equation $y = t(x + 1)$, where $t = h/(g + 1)$,

$$g = \frac{1 - t^2}{1 + t^2},$$    (3)

and

$$h = \frac{2t}{1 + t^2}.$$    (4)

*Proof.* The discussion preceding the lemma shows that the equation of the line $\ell$ is as advertised.

Since $(g, h)$ lies on the unit circle, we have $1 = g^2 + h^2$; since $(g, h)$ lies on the line $\ell$, we have $h = t(g + 1)$. Hence,

$$1 = g^2 + h^2 = g^2 + t^2(g + 1)^2 = g^2 + t^2 g^2 + 2t^2 g + t^2.$$

Collecting terms, we have a quadratic equation in $g$:

$$(t^2 + 1)g^2 + 2t^2 g + (t^2 - 1) = 0.$$

Since the line $\ell$ meets the unit circle in two points, $(-1, 0)$ and $(g, h)$, it follows that the roots of $(t^2 + 1)x^2 + 2t^2 x + t^2 - 1$ are $x = -1$ and $x = g$. We are going to use the observation that if $r$ and $s$ are roots of $ax^2 + bx + c$, then

$$ax^2 + bx + c = a(x - r)(x - g)$$

and $c = arg$. As $r = -1$ here, we have $t^2 - 1 = -(t^2 + 1)g$, and so

$$g = \frac{1 - t^2}{1 + t^2}.$$

We now solve for $h$. Because $(g, h)$ lies on the line $\ell$, Eq. (2) gives

$$\begin{aligned}
h &= t(g + 1) \\
&= t\left[\left(\frac{1 - t^2}{1 + t^2}\right) + 1\right] \\
&= \frac{t(1 - t^2 + 1 + t^2)}{1 + t^2} \\
&= \frac{2t}{1 + t^2}. \quad \bullet
\end{aligned}$$

**Lemma 2.10.** Let $(g, h)$ be a point on the unit circle other than $(-1, 0)$. If $t$ is the slope of the line joining $(-1, 0)$ and $(g, h)$, then $t$ is a rational number if and only if $(g, h)$ is a rational point.

*Proof.* If $(g, h)$ is a rational point, then both $g$ and $h$ are rational numbers. It is now plain that $t = h/(g + 1)$ is also a rational number.

Conversely, assume that $t$ is a rational number. Now

$$g = \frac{1 - t^2}{1 + t^2},$$

by Eq. (3), and this shows that $g$ is a rational number; also,

$$h = \frac{2t}{1 + t^2},$$

by Eq. (4), so that $t$ a rational number implies that both $g$ and $h$ are rational numbers; that is, $(g, h)$ is a rational point.   •

We are going to use the work above to discover the formula stated in the preceding section, and also to show that the formula gives essentially all Pythagorean triples. But first we encode the side conditions that $a, b, c$ are positive integers and that $a < b$ into geometry.

**Lemma 2.11.** If $(a, b, c)$ is a Pythagorean triple, then the point $(a/c, b/c)$ is a rational point on the arc of the unit circle in the first quadrant above the line $L$ having equation $y = x$. Moreover,

$$\left( \frac{a}{c}, \frac{b}{c} \right) = \left( \frac{1 - t^2}{1 + t^2}, \frac{2t}{1 + t^2} \right),$$

where $.414 \approx \sqrt{2} - 1 < t < 1$.

*Proof.* A point $(x, y)$ in the plane with both coordinates positive lies in the first quadrant. If $x < y$, then $(x, y)$ lies above the line $L$, for $(x, x)$ lies on $L$, and $x < y$ shows that $(x, y)$ is higher on the vertical line through $(x, 0)$ than $(x, x)$ is.

That $(a/c, b/c) = ([1 - t^2]/[1 + t^2], 2t/[1 + t^2])$ follows from Lemma 2.9, and so we need only check the inequalities. As $t$ is the slope of the line joining $(-1, 0)$ to points on the unit circle, it suffices to compute the slopes of these lines to the points $(0, 1)$ and $(1/\sqrt{2}, 1/\sqrt{2})$. For the first point, the slope $t = 1$; for the second point $(0, 1)$, the slope is

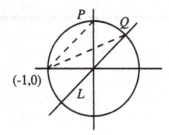

Figure 2.33

$$t = \frac{\frac{1}{\sqrt{2}}}{1 + \frac{1}{\sqrt{2}}} = \frac{1}{1 + \sqrt{2}} = \sqrt{2} - 1.$$

Therefore, if $t$ is the slope of a line through $(-1, 0)$ and some point on the arc of the unit circle joining $(\frac{1}{\sqrt{2}}, \frac{1}{\sqrt{2}})$ and $(0, 1)$, then

$$.414 \approx \sqrt{2} - 1 < t < 1. \quad \bullet$$

**Theorem 2.12**. Every Pythagorean triple $(a, b, c)$ is similar to a Pythagorean triple of the form

$$(q^2 - p^2, 2qp, q^2 + p^2),$$

where $q > p$ are positive integers with

$$.414 \approx \sqrt{2} - 1 < \frac{p}{q} < 1.$$

*Proof.*  Given a Pythagorean triple $(a, b, c)$, we obtain a rational point $(a/c, b/c)$ on the unit circle. If $t$ is the slope of the line joining $(-1, 0)$ and $(a/c, b/c)$, then Lemma 2.11 gives

$$\frac{a}{c} = \frac{1 - t^2}{1 + t^2}$$

and

$$\frac{b}{c} = \frac{2t}{1 + t^2},$$

where $\sqrt{2} - 1 < t < 1$. By Lemma 2.10, $t$ is rational, say, $t = p/q$. Hence,

$$\frac{a}{c} = \frac{1 - t^2}{1 + t^2}$$

$$= \frac{1 - (p/q)^2}{1 + (p/q)^2}$$

$$= \frac{q^2 - p^2}{q^2 + p^2},$$

and

$$\frac{b}{c} = \frac{2t}{1 + t^2}$$

$$= \frac{2p/q}{1 + (p/q)^2}$$

$$= \frac{2qp}{q^2 + p^2}.$$

Now $(q^2 - p^2, 2qp, q^2 + p^2)$ is a Pythagorean triple if $q^2 - p^2 < 2qp$. But $a/c < b/c$, by Lemma 2.11, so that

$$\frac{q^2 - p^2}{q^2 + p^2} < \frac{2qp}{q^2 + p^2},$$

and, hence, $q^2 - p^2 < 2qp$.

It remains to show that $(q^2 - p^2, 2qp, q^2 + p^2)$ is similar to $(a, b, c)$. The equations

$$\frac{a}{c} = \frac{q^2 - p^2}{q^2 + p^2} \quad \text{and} \quad \frac{b}{c} = \frac{2qp}{q^2 + p^2}$$

yield, after cross multiplying, the equations

$$a(q^2 + p^2) = c(q^2 - p^2) \quad \text{and} \quad b(q^2 + p^2) = c(2qp).$$

Of course, $c(q^2 + p^2) = (q^2 + p^2)c$, and so the Pythagorean triples are similar: in the notation in the definition of similarity, $\ell = q^2 + p^2$ and $m = c$.  ●

The strategy of Diophantus is quite elegant. The problem of determining all Pythagorean triples is reduced from finding three unknowns, namely, $a, b$, and $c$, to two unknowns, namely, $g = a/c$ and $h = b/c$, to one unknown, namely $t = p/q$. In effect, all Pythagorean triples are parametrized by $t$; that is, as $t$ varies over rational numbers, the formulas involving $t$ vary over Pythagorean triples.

Not every Pythagorean triple $(a, b, c)$ arises from the formula [nor does the theorem say that it does; the theorem asserts only that there is some Pythagorean triple arising from the formula that is similar to $(a, b, c)$]. For example, let us show that $(9, 12, 15)$ is not of the form $(q^2 - p^2, 2qp, q^2 + p^2)$. Since the leg 9 is odd, the even leg 12 must be $2qp$, so that $qp = 6$; hence, $q = 3$ and $p = 2$, or $q = 6$ and $p = 1$. Since $p/q = 1/6 \approx .166 < .414 \approx \sqrt{2} - 1$, we can ignore it [the corresponding Pythagorean triple, however, is $(12, 35, 37)$]. The values $q = 3$ and $p = 2$ give the Pythagorean triple $(5, 12, 13)$, not $(9, 12, 15)$. On the other hand, $(9, 12, 15)$ is similar to $(3, 4, 5)$, and $(3, 4, 5)$ does arise from the formula with $q = 2$ and $p = 1$. As a consequence, the $q > p$ method used in the last section (for testing whether an alleged Pythagorean triple $(a, b, c)$ really is one) does not always work. If one can find integers $q > p$ for which $c = q^2 + p^2$, etc., then $(a, b, c)$ is a Pythagorean triple. However, if integers $q$ and $p$ cannot be found, the method is inconclusive: $(a, b, c)$ might still be a Pythagorean triple. For example, we have just seen that the method does not decide whether or not $(9, 12, 15)$ is a Pythagorean triple.

In the 1630's, P. Fermat (1601–1665) wrote in the margin of his copy of a book by Diophantus[11]:

> . . . it is impossible for a cube to be written as a sum of two cubes or a fourth power to be written as a sum of two fourth powers or, in general, for any number which is a power greater than the second to be written as a sum of two like powers. I have discovered a truly marvelous demonstration of this proposition which this margin is too narrow to contain.

Fermat never returned (in print) to this problem, except for his proof for $n = 4$. The statement, "For every $n > 2$, there are no positive integers

---

[11]Not everyone appreciated Diophantus. Another copy of Diophantus, not Fermat's, but from about the same time, also contains a note in its margin. Referring to the version of Theorem 2.12, it says, "May your soul, Diophantus, be with the Devil, because of the difficulty of your theorems and, above all, the difficulty of this one!"

THE METHOD OF DIOPHANTUS     71

$a, b, c$ with $a^n + b^n = c^n$" was (waggishly) called "Fermat's Last Theorem." Fermat's Last Theorem became a famous problem, and many important techniques were developed in unsuccessful attempts to prove it. Indeed, these techniques have turned out to be more important than the problem itself. In 1908, a prize of 100,000 German marks was offered in the will of Paul Wolfskehl for a proof. At last, in 1995, Andrew Wiles, with Richard Taylor, succeeded in finding a proof that had eluded mathematicians for more than 350 years (the proof is very sophisticated, and it is difficult to believe that Fermat himself actually had a correct proof.)

### Exercises

**2.29.** In Lemma 2.10, find the values of $g$ and $h$ for $t = \frac{2}{3}, \frac{3}{4}, \frac{3}{5},$ and $\frac{4}{5}$.

**2.30.** Find $q$ and $p$ in Theorem 2.12 for each of the following Pythagorean triples.
(i) $(7, 24, 25)$.
(ii) $(8, 15, 17)$. (Hint: Since the middle leg is odd, consider the similar triple $(16, 30, 34)$.)
(iii) $(129396, 261547, 291805)$.

**2.31.** Show that the same number can occur as a leg in two nonsimilar Pythagorean triples. (Hint: Try 8.)

**2.32.** Show that there are distinct Pythagorean triples $(a, b, c)$ and $(\alpha, \beta, c)$ having the same $c$.

**2.33.** Show that every integer $n \geq 3$ occurs as a leg of some Pythagorean triple. Hint: Let $n$ be even, say, $n = 2k$. If $n = 4$, then $(3, 4, 5)$ works; if $n \geq 6$, then $n < k^2 - 1$, and
$$(n, k^2 - 1, k^2 + 1)$$
is a Pythagorean triple. If $n$ is odd, say, $n = 2k + 1$, then
$$(n, 2k(k + 1), 2k^2 + 2k + 1)$$
is a Pythagorean triple.

## TRIGONOMETRY

Since you are now studying geometry and trigonometry, I will give you a problem. A ship sails the ocean. It left Boston with a cargo of wool. It grosses 200 tons. It is bound for Le Havre. The mainmast is broken, the cabin boy is on deck, there are 12 passengers aboard, the wind is blowing East-North-East, the clock points to a quarter past three in the afternoon. It is the month of May. How old is the captain?

*Gustave Flaubert*

Trigonometric functions first arose over 3000 years ago in relation to right triangles (the word *trigonometry* means "triangle measure").

Let $\triangle ABC$ be a right triangle with side lengths $a$, $b$ and $c$, and let $\alpha$ be one of its acute angles.

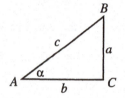

Figure 2.34

Using the notation in Figure 2.34, we define the six trigonometric functions: sine, cosine, tangent, cotangent, secant, and cosecant.

$$\sin \alpha = a/c,$$
$$\cos \alpha = b/c,$$
$$\tan \alpha = a/b,$$
$$\cot \alpha = b/a,$$
$$\sec \alpha = c/b,$$
$$\csc \alpha = c/a.$$

At this point, these functions are defined only for acute angles $\alpha$. There is an obvious question: Do these functions depend only on $\alpha$, or do they depend upon the choice of right triangle as well? A second right triangle $\triangle A'B'C'$ with angle $\alpha$ and sides of lengths $a'$, $b'$ and $c'$ must be similar to

$\triangle ABC$, and so $a'/c' = a/c$; that is, both triangles give the same value for $\sin\alpha$. The same argument shows that the other five trigonometric functions do not depend on the choice of right triangle either. If the right triangle is chosen with $c = 1$, then it follows that $\sin\alpha = a$ and $\cos\alpha = b$.

Let us see how trigonometry can help solve practical measuring problems.

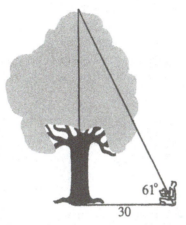

Figure 2.35

A man sits 30 feet from the base of a tree, looking up at the treetop (see Figure 2.35). Using a protractor-like device, he finds that his line of sight makes an angle $\alpha = 61°$. Approximately how tall is the tree?

If $h$ is the height of the tree, then $\tan 61° = h/30$, or $h = 30\tan 61°$. Since $\tan 61° \approx 1.80$, the tree is approximately $30 \times 1.80 = 54$ feet tall.

We see, therefore, that trigonometry is quite practical *if* one has a table of values. The first accurate trigonometric table is the *Almagest* of Ptolemy (Claudius Ptolemaeus, ca.100–ca.178), where one finds values for increments of half-degrees of angles. Ptolemy's method is based on plane geometry and the addition formulas [for $\sin(\alpha + \beta)$ and $\cos(\alpha + \beta)$], as well as on work of Hipparchus 300 years earlier.

***Theorem 2.13***. For every acute angle $\alpha$, we have

$$\sin^2\alpha + \cos^2\alpha = 1.$$

*Proof.* Choose a right triangle having an angle $\alpha$ and sides of lengths $a$, $b$, and $c$, so that $\sin \alpha = a/c$ and $\cos \alpha = b/c$. By the Pythagorean theorem, $a^2 + b^2 = c^2$. Dividing by $c^2$ gives $(a/c)^2 + (b/c)^2 = 1$, as desired.    •

One of my teachers, S. Mac Lane, defined a mathematician as a person who, upon seeing something a second time, hears a bell ring. Surely, Theorem 2.13 reminds you of the unit circle with equation $x^2 + y^2 = 1$. The Pythagorean theorem is of fundamental importance because it gives the distance formula between two points. This leads to the equation $x^2 + y^2 = 1$ describing the unit circle [as all points $(x, y)$ whose distance to the origin is 1], and this equation gives the basic equation of trigonometry: $\sin^2 \alpha + \cos^2 \alpha = 1$.

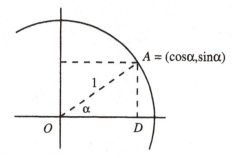

Figure 2.36

Choose a point $A = (x, y)$ in the first quadrant lying on the unit circle. The line $OA$ makes an angle $\alpha$ with the $x$-axis. If $D$ denotes the foot of the perpendicular from $A$, then $y = |AD|$ and $x = |OD|$. Because the radius $OA$ has length 1, we have

$$x = |OD| = \frac{|OD|}{|OA|} = \cos \alpha$$

$$y = |AD| = \frac{|AD|}{|OA|} = \sin \alpha.$$

Therefore, $(x, y) = (\cos \alpha, \sin \alpha)$. From this point of view, we see that the discussion of Pythagorean triples is a description of all those angles $\alpha$ for which both $\sin \alpha$ and $\cos \alpha$ are rational. Note that it is possible for one of these numbers to be rational and the other irrational; for example, $\sin 30° = \frac{1}{2}$ while

$\cos 30° = \sqrt{3}/2$. In the last chapter, we shall display angles $\alpha$ for which both $\sin \alpha$ and $\cos \alpha$ are irrational.

At the moment, the trigonometric functions make sense only for acute angles, for we defined them in terms of the angles in a right triangle. It is now a simple matter, however, to extend the definitions of these functions to all angles $\alpha$, be they obtuse or acute, negative (measured clockwise) or positive (measured counter-clockwise). Any angle $\alpha$ can be drawn so that its vertex is at the origin and one of its sides is the positive $x$-axis. Let $A$ be the point of intersection of the other side of the angle with the unit circle, and define $\cos \alpha$ and $\sin \alpha$ as the coordinates $(x, y)$ of $A$. It follows at once that Theorem 2.13 holds for all, not necessarily acute, angles. One can now check familiar identities such as $\cos(-\alpha) = \cos \alpha$ and $\sin(-\alpha) = -\sin \alpha$.

There are two common ways of measuring angles. The most popular one, dividing the circle into 360 degrees, goes back to Babylonian times. We do not know why they chose 360 degrees and not 100 or some other number. Perhaps the Babylonians wanted a number with many factors that is close to the number of days in a year. The most popular angle measure in science uses radians. Although the notion of radian measure probably goes back to the Greeks, the *word* "radian" first appeared in print in 1873 (used by James Thomson, the brother of Lord Kelvin). Any point $A$ on the unit circle, together with the origin, determines a line which, with the $x$-axis, determines an angle. The radian measure of the angle is defined as the arclength from $(0, 1)$ to $A$ (measured counterclockwise). The arclength of a complete circuit is the circumference $2\pi$, but without the circumference formula, one can neither compute the radian measure of the whole circle nor, indeed, the radian measure of any nonzero angle.

Let us see where the terms *tangent* and *secant* come from. In Figure 2.37, $C = (1, 0)$ and $BC$ is tangent to the circle at $C$ (thus, $BC$ is perpendicular to $OC$). Since $\triangle OAD$ and $\triangle OBC$ are similar, we have

$$\frac{|BC|}{|AD|} = \frac{|OC|}{|OD|}.$$

But $|AD| = \sin \alpha$, $|OC| = 1$ (it is a radius), and $|OD| = \cos \alpha$. Hence

$$|BC| = \frac{\sin \alpha}{\cos \alpha} = \tan \alpha.$$

Thus, $\tan \alpha$ is the length of the tangent $BC$ (for the Latin word *tangere* means "to touch," and a *tangent* is a line which *touches* the circle in only one point).

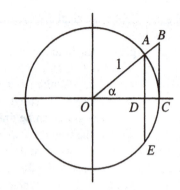

Figure 2.37

Similarity also gives the proportion

$$\frac{|OB|}{|OA|} = \frac{|OC|}{|OD|}.$$

But $|OA| = 1 = |OC|$ (they are radii) and $|OD| = \cos\alpha$, so that

$$|OB| = \frac{1}{\cos\alpha} = \sec\alpha.$$

Thus, $\sec\alpha$ is the length of the secant $OB$ (for the Latin word *secare* means "to cut," and a *secant* is a line which *cuts* a circle). The *complement* of an acute angle $\alpha$ is $90° - \alpha$, and so the name *cosine* arises from that of sine because of the identity $\cos\alpha = \sin(90° - \alpha)$.

The reason for the term *sine* is more amusing. We see in Figure 2.37 that

$$\sin\alpha = |AD| = \tfrac{1}{2}|AE|;$$

that is, $\sin\alpha$ is half the length of the chord $AE$. The fifth century Indian mathematician Aryabhata called the sine *ardha-jya* (half chord) in Sanskrit; this term was later abbreviated to *jya*. A few centuries later, books in Arabic transliterated *jya* as *jiba*. In Arabic script, there are letters and diacritical marks; roughly speaking, the letters correspond to our consonants, while the diacritical marks correspond to our vowels. It is customary to suppress diacritical marks in writing; for example, the Arabic version of *jiba* is written *jb* (using

Arabic characters, of course). Now *jiba*, having no other meaning in Arabic, eventually evolved into *jaib*, which is an Arabic word, meaning "bosom of a dress" (a fine word, but having absolutely nothing to do with half-chord). Finally, Gherardo of Cremona, ca. 1150, translated *jaib* into its Latin equivalent *sinus* (this Latin word has many meanings, all related to curved shapes). And this is why sine is so called, for sine means bosom!

We now rephrase Lemma 2.9.

***Theorem 2.14.*** If $0 \leq \theta < 360°$ and $\theta \neq 180°$, one can choose $t$ so that

$$\cos\theta = \frac{1 - t^2}{1 + t^2} \quad \text{and} \quad \sin\theta = \frac{2t}{1 + t^2}.$$

*Proof.* In Lemma 2.9, each point $(g, h)$ on the unit circle, other than $(-1, 0)$, was parametrized by

$$g = \frac{1 - t^2}{1 + t^2} \quad \text{and} \quad h = \frac{2t}{1 + t^2}.$$

We now recognize that $g = \cos\theta$ and $h = \sin\theta$.    •

Many trigonometric identities can be proved using Theorem 2.14: $\cos\theta = (1 - t^2)/(1 + t^2)$ and $\sin\theta = 2t/(1 + t^2)$. First, rewrite everything in terms of $\cos\theta$ and $\sin\theta$; second, replace each occurrence of $\cos\theta$ and $\sin\theta$ by its expression as a rational function of $t$; third, check whether the resulting equation in $t$ is valid (all these steps are mechanical and require no ingenuity). For example, consider the identity:

$$\frac{1 + \cos\theta + \sin\theta}{1 + \cos\theta - \sin\theta} = \sec\theta + \tan\theta.$$

After applying Theorem 2.14, the left side is

$$\frac{1 + (1 - t^2)/(1 + t^2) + 2t/(1 + t^2)}{1 + (1 - t^2)/(1 + t^2) - 2t/(1 + t^2)},$$

which simplifies to $(1 + t)/(1 - t)$, while the right side is

$$\frac{1 + t^2}{1 - t^2} + \frac{2t/(1 + t^2)}{(1 - t^2)/(1 + t^2)},$$

which also simplifies to the same thing. The reader should try to prove this identity the "old-fashioned way." Here is another "tricky" identity that surrenders easily to the rational function substitution.

$$1 + \csc \theta = \frac{\cos \theta \cot \theta}{1 - \sin \theta}.$$

We quote [Silverman-Tate]:

> If they had told you this in high school, the whole business of trigonometric identities would have become a trivial exercise in algebra!

**Theorem 2.15 ( Law of Cosines).** Let $\triangle ABC$ be a triangle with angles $\alpha$, $\beta$, $\gamma$ at $A$, $B$, and $C$, respectively. If $a$, $b$, and $c$ denote the lengths of sides opposite $A$, $B$, $C$, then

$$a^2 = b^2 + c^2 - 2bc \cos \alpha.$$

*Proof.* Assume that $\alpha$ is acute. In Figure 2.38a, set $p = |AD|$ and $q = |DB|$, so that

$$q = c - p.$$

The Pythagorean theorem gives

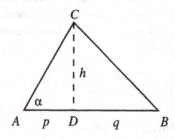

Figure 2.38a

$$a^2 = h^2 + q^2 \quad \text{and} \quad h^2 = b^2 - p^2,$$

while $\cos \alpha = p/b$ gives

$$p = b \cos \alpha.$$

We now compute:

$$a^2 = h^2 + q^2$$
$$= b^2 - p^2 + q^2$$
$$= b^2 - p^2 + (c - p)^2$$
$$= b^2 - p^2 + c^2 - 2cp + p^2$$
$$= b^2 + c^2 - 2cp$$
$$= b^2 + c^2 - 2bc \cos \alpha.$$

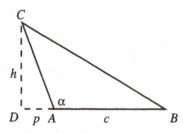

Figure 2.38b

As in the proof of Theorem 2.1, we must still consider the case when $\alpha$ is obtuse, as drawn in Figure 2.38b. This proof is left to the reader; it is very similar to the acute case, although it will need the fact that $\cos(180° - \alpha) = -\cos \alpha$. •

The following theorem was proved by Heron of Alexandria in the first century.

**Theorem 2.16 (Heron).** *If $\triangle ABC$ has sides of lengths $a$, $b$, and $c$, and if the* **semiperimeter** $s = \frac{1}{2}(a + b + c)$, *then*

$$\text{area}(\triangle ABC)^2 = s(s - a)(s - b)(s - c).$$

*Proof.* We refer the reader to Figure 2.38a. Area$(\triangle ABC) = \frac{1}{2}hc = \frac{1}{2}bc \sin \alpha$, so that

$$\text{area}(\triangle ABC)^2 = (\tfrac{1}{2}bc \sin \alpha)^2 = \tfrac{1}{4}b^2c^2 \sin^2 \alpha.$$

The law of cosines gives

$$a^2 = b^2 + c^2 - 2bc \cos \alpha,$$

so that

$$\cos \alpha = \frac{b^2 + c^2 - a^2}{2bc}.$$

Hence,

$$
\begin{aligned}
\text{area}(\triangle ABC)^2 &= \tfrac{1}{4}b^2 c^2 (1 - \cos^2 \alpha) \\
&= \tfrac{1}{4}b^2 c^2 \left[ 1 - ([b^2 + c^2 - a^2]/2bc)^2 \right] \\
&= \tfrac{1}{16} \left[ 4b^2 c^2 - (b^2 + c^2 - a^2)^2 \right].
\end{aligned}
$$

We are essentially finished, for this last formula gives area($\triangle ABC$) in terms of $a$, $b$, and $c$. The rest of the proof is merely cosmetic.

Using the identity $x^2 - y^2 = (x - y)(x + y)$ at appropriate times, we have

$$
\begin{aligned}
\text{area}(\triangle ABC)^2 &= \tfrac{1}{16}[4b^2 c^2 - (b^2 + c^2 - a^2)^2] \\
&= \tfrac{1}{16}[2bc - (b^2 + c^2 - a^2)][2bc + (b^2 + c^2 - a^2)] \\
&= \tfrac{1}{16}[2bc - b^2 - c^2 + a^2][b^2 + 2bc + c^2 - a^2] \\
&= \tfrac{1}{16}[a^2 - (b - c)^2][(b + c)^2 - a^2] \\
&= \tfrac{1}{16}(a - b + c)(a + b - c)(b + c - a)(b + c + a).
\end{aligned}
$$

Because

$$a - b + c = 2(s - b), \qquad a + b - c = 2(s - c),$$
$$b + c - a = 2(s - a), \qquad b + c + a = 2s,$$

substitution yields the desired result.   •

An *isoperimetric problem* (*isoperimetric* means "same perimeter" in Greek; *perimeter* means "measure around") asks, of all figures of a certain shape and whose perimeter is of a given length, which of them has the largest area. We are now going to solve the isoperimetric problem for triangles.

**Theorem 2.17.** Of all triangles having perimeter of a given length $p = 2s$, the equilateral triangle has the largest area, namely, $s^2/\sqrt{27}$.

*Proof.* Let $T$ be a triangle with sides of lengths $a$, $b$, $c$ and perimeter $2s = a + b + c$. By Heron's theorem, if $A = \text{area}(T)$, then $A^2 = s(s-a)(s-b)(s-c)$, so that

$$A^2/s = (s-a)(s-b)(s-c).$$

By Theorem 1.11, the inequality of the means,

$$\left[\frac{(s-a)+(s-b)+(s-c)}{3}\right]^3 \geq (s-a)(s-b)(s-c) = A^2/s,$$

with equality holding if and only if $s - a = s - b = s - c$; that is, $a = b = c$. The term in brackets is $(3s - 2s)/3 = s/3$, so that $s^3/27 \geq A^2/s$. Hence, $s^4/27 \geq A^2$, and $s^2/\sqrt{27} \geq A$, with equality if and only if $T$ is equilateral. Therefore, the equilateral triangle has the largest area, which is $s^2/\sqrt{27}$. •

It follows that an equilateral triangle with sides of length $a$ has area $s^2/\sqrt{27} = a^2\sqrt{3}/4$, which is the formula given in Exercise 2.20(ii).

Theorem 2.17 can also be proved by calculus. For given $s > 0$, define

$$f(x, y, z) = (s-x)(s-y)(s-z)$$

[we are using Heron's formula]. The problem is to maximize $f(x, y, z)$ over all points $(x, y, z)$ in the first octant subject to the constraint $x + y + z = 2s$. The method of Lagrange multipliers leads to only one constrained maximum, namely, $(\frac{2}{3}s, \frac{2}{3}s, \frac{2}{3}s)$, giving the result. Our first proof, however, uses much less machinery.

The next theorem, important for trigonometry, is also important for applications of complex numbers, as we shall see in Chapter 4.

**Theorem 2.18 (Addition Theorem).** For all angles $x$ and $y$,

$$\cos(x + y) = \cos x \cos y - \sin x \sin y$$

and

$$\sin(x + y) = \cos x \sin y + \sin x \cos y.$$

*Proof.* Let $\alpha$ and $\beta$ be angles. In Figure 2.39, $L = (1, 0)$, $A = (\cos \alpha, \sin \alpha)$, $B = (\cos \beta, \sin \beta)$, and $C = (\cos \gamma, \sin \gamma)$, where $\gamma = \alpha + \beta$. Now $\triangle AOL \cong \triangle COB$, by "side-angle-side," so that $|CB| = |AL|$; the distance formula[12] $d(C, B)^2 = d(A, L)^2$ gives

---

[12]Recall that the distance between points $(x, y)$ and $(a, b)$ is $\sqrt{(x-a)^2 + (y-b)^2}$.

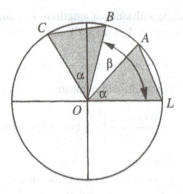

Figure 2.39

$$(\cos \gamma - \cos \beta)^2 + (\sin \gamma - \sin \beta)^2 = (\cos \alpha - 1)^2 + \sin^2 \alpha.$$

Using Theorem 2.13, the left side is:

$$\cos^2 \gamma - 2 \cos \gamma \cos \beta + \cos^2 \beta + \sin^2 \gamma - 2 \sin \gamma \sin \beta + \sin^2 \beta$$

$$= 2 - 2(\cos \gamma \cos \beta + \sin \gamma \sin \beta),$$

whereas the right side is

$$\cos^2 \alpha - 2 \cos \alpha + 1 + \sin^2 \alpha = 2 - 2 \cos \alpha.$$

Therefore, for every $\alpha$ and $\beta$,

$$\cos \alpha = \cos \gamma \cos \beta + \sin \gamma \sin \beta$$
$$= \cos(\alpha + \beta) \cos \beta + \sin(\alpha + \beta) \sin \beta.$$

Setting $\alpha = x - \beta$, we have

$$\cos(x - \beta) = \cos x \cos \beta + \sin x \sin \beta,$$

and setting $\beta = -y$, we have

$$\cos(x + y) = \cos x \cos y - \sin x \sin y.$$

To prove the addition formula for sine, observe first that

$$\cos(90° + z) = \cos 90° \cos z - \sin 90° \sin z = -\sin z. \qquad (5)$$

Setting $z = x + 90°$ and $y = -90°$, we have

$$\begin{aligned}
\cos z &= \cos(90° + z - 90°) \\
&= \cos(90° + z)\cos(90°) - \sin(90° + z)\sin(-90°) \\
&= -\sin(90° + z)\sin(-90°) = \sin(90° + z). \qquad (6)
\end{aligned}$$

Therefore, setting $z = x + y$, we have

$$\begin{aligned}
\sin(x + y) &= -\cos(90° + [x + y]) \\
&= -\cos([90° + x] + y) \\
&= -[\cos(90° + x)\cos y - \sin(90° + x)\sin y] \\
&= -\cos(90° + x)\cos y + \sin(90° + x)\sin y \\
&= \sin x \cos y + \cos x \sin y,
\end{aligned}$$

by Eqs. (5) and (6).  •

## Exercises

**2.34.** Use Heron's formula to solve Exercise 2.15(i).

**2.35.** Let $P = (1, 1)$, $Q = (2, 3)$, and $R = (-2, 2)$ be three points in the plane.
(i) Show that $P$, $Q$, and $R$ are not collinear.
(ii) Find area$(\triangle PQR)$.

**2.36.** To find the distance between two points $A$ and $C$ on opposite sides of a river, a distance of 100 feet is paced off on one side of the river from $A$ to a point $B$ with $AB$ perpendicular to $BC$. The angle $\angle CAB$ is measured as 60°. Find the distance $|AC|$. (See Figure 2.40.)

**2.37.** A pole 30 feet tall casts a shadow 50 feet long. Find the approximate angle of elevation of the sun.

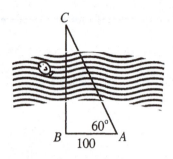

Figure 2.40

**2.38.** New Orleans, Louisiana is due south of Madison, Wisconsin. If the latitude of New Orleans is 30° N, the latitude of Madison is 43° N, and the radius of the earth is 4000 miles, what is the distance between the two cities?

Answer: 907.6 miles.

**2.39.** Assuming the fact that the arc and the chord subtended by a small central angle in a circle are approximately the same length, compute the diameter of the Sun using the facts that the distance from the Earth to the Sun is about 93,000,000 miles and that the Sun as seen from Earth subtends an angle of 0.53 degrees.

Answer: 430,000 miles.

**2.40.** If a regular decagon whose sides are 7 inches long is inscribed in a circle, what is the radius of the circle?    Answer: 11.32 inches.

**2.41.** Let $\alpha$ be an angle and let $A = (\cos\alpha, \sin\alpha)$. Show that if $m$ is the slope of the line $OA$ (where $O$ denotes the origin), then $m = \tan\alpha$.

**2.42.** Let $(g, h)$ be a point in the first quadrant lying on the arc of the unit circle, and let the line $\ell$ joining $(g, h)$ and $(-1, 0)$ have slope $t$. Prove that if the line $L$ joining $(h, g)$ and $(-1, 0)$ has slope $T$, then $T = (1 - t)/(1 + t)$. [Hint: Use Eqs. (2), (3), and (4).]

**2.43.** Prove that

$$\frac{1}{\csc\theta - \cot\theta} - \frac{1}{\csc\theta + \cot\theta} = 2\cot\theta.$$

**2.44.** Prove that $\cot^4\theta + \cot^2\theta = \csc^4\theta - \csc^2\theta$.

**2.45.** Of all the triangles with a given perimeter $p$ and with one side of a given length $a$, prove that the triangle having the maximal area is isosceles.

**2.46.** Prove, for every triangle $\triangle$ , that

$$\text{area}(\triangle) \le \frac{p^2}{12\sqrt{3}},$$

where $p$ is the perimeter of $\triangle$. Can equality hold?

**2.47.** Prove the *Law of Sines:* In a triangle $\triangle ABC$ with angles $\alpha$, $\beta$, and $\gamma$ opposite sides of lengths $a$, $b$, and $c$, respectively,

$$\frac{\sin\alpha}{a} = \frac{\sin\beta}{b} = \frac{\sin\gamma}{c}.$$

(Hint: Because $\text{area}(\triangle ABC) = \frac{1}{2}bh = \frac{1}{2}bc\sin\alpha$, we have

$$\frac{\sin\alpha}{a} = \frac{2\,\text{area}(\triangle ABC)}{abc.}$$

Show that $\sin\beta/b$ and $\sin\gamma/c$ also equal this number.)

**2.48.** For any two angles $\alpha$ and $\beta$, prove

$$\cos[\tfrac{1}{2}(\alpha + \beta)]\cos[\tfrac{1}{2}(\alpha - \beta)] = \tfrac{1}{2}(\cos\alpha + \cos\beta).$$

**2.49.** Prove that $\cos 20° \cos 40° \cos 80° = \frac{1}{8}$. Note that this is an equation, not an approximation, and so calculators cannot prove it. (Hint: Use Exercise 2.48.)

**2.50.** (i) Prove that

$$\cos x \cos 2x \cos 4x \cdots \cos 2^n x = \frac{\sin 2^{n+1}x}{2^{n+1}\sin x}.$$

(Hint: Multiply the left side by $\sin x / \sin x$.)
(ii) Use part (i) to give a second proof of Exercise 2.49.

**2.51.** Prove that $2\cos(n + 1)x = (2\cos x)(2\cos nx) - 2\cos(n - 1)x$.

## INTEGRATION

Common integration is only the memory of differentiation.

*A. De Morgan*

The definite integral $\int_a^b f(x)\,dx$ has a very complicated definition: it is a limit, over all partitions $a = x_0 < x_1 < \cdots < x_n = b$ of the interval $[a, b]$, of sums $\sum_{i=1}^n f(\theta_i)(x_i - x_{i-1})$, where $\theta_i$ lies in the interval $[x_{i-1}, x_i]$. If one could not evaluate integrals for interesting classes of functions, then this notion would be useless. Archimedes[13] would be pleased to hear that the area under the arch of $\sin x$ over the interval $[0, \pi]$ is 2; he would not be pleased at being told that this area is $\lim \sum_{i=1}^n \sin(\theta_i)(x_i - x_{i-1})$. The fundamental theorem of calculus: if $F(x) = \int_a^x f(t)\,dt$, then $F'(x) = f(x)$ and $\int_a^b f(t)\,dt = F(b) - F(a)$, evaluates definite integrals if one remembers derivatives. We know the formula $(x^{n+1})' = (n + 1)x^n$, and this is what allows us to integrate polynomials. As another example, the formula $(-\cos x)' = \sin x$ is what allows us to please Archimedes.

I realize one should write $\int f(x)\,dx = F(x) + c$ if $F(x) = \int_a^x f(t)\,dt$, but I usually forget to write the constant $c$. But everyone knows this. There is a story of two friends arguing, David saying that everyone knows calculus, Jonathan denying it. They went into a restaurant and, when Jonathan was away from the table, David called the waitress. "The next time you come to this table, I'll ask you a question. Will you please answer, '$x$ squared over 2'?" The argument continued after Jonathan returned. "I'll show you that everyone knows calculus," David said. He called the waitress and asked her, "Do you know the indefinite integral of $x\,dx$?" "Yes, I do," she replied, "$x$ squared over 2 plus a constant."

After polynomials, the next most natural class of functions consists of **rational functions** $g(x)/h(x)$, where $g(x)$ and $h(x)$ are polynomials. Now $\int [g(x)/h(x)]\,dx$ can be evaluated by the method of partial fractions. There is an algebraic way to write $g(x)/h(x)$ as the sum of a polynomial and terms of the form

$$\frac{A}{(ax + b)^n} \quad \text{and} \quad \frac{Bx + C}{(ax^2 + bx + c)^m},$$

---

[13] Archimedes discovered many formulas without integration: for example, the volume of a sphere; volumes of segments of surfaces of revolution of conic sections; areas of segments of parabolas.

and the method of substitution allows one to integrate either of these forms. Recall that the ultimate answers can involve logs and arctans.

Next, one considers the trigonometric functions. Integrating tangent needs a bit of thought, but integrating secant is not at all straightforward. Many calculus texts use some *ad hoc* trick, whereas others use the substitution $t = \tan(\theta/2)$, which magically works. We are now going to see that the formulas that arose when we studied Pythagorean triples explain the magic. In Figure 2.41, $A = (-1, 0)$ and $B = (\cos\theta, \sin\theta)$. We claim that $\angle OAB = \theta/2$. After all, $\theta$ is an exterior angle of $\triangle AOB$, so that $\theta = \angle OAB + \angle OBA$. Now $\angle OAB = \angle OBA$, because $\triangle AOB$ is an isosceles triangle (for two sides are radii), and hence $\theta = 2\angle OAB$.

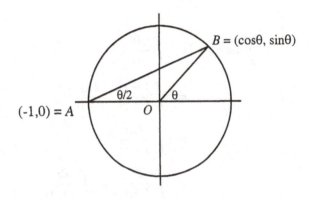

Figure 2.41

It follows from Exercise 2.52 that

$$\frac{\sin\theta}{\cos\theta + 1} = \text{slope } AB = \tan(\theta/2). \qquad (7)$$

If we set

$$t = \tan(\theta/2),$$

then

$$\theta = 2\arctan t \quad \text{and} \quad d\theta = \frac{2dt}{1 + t^2}.$$

We proved, in Theorem 2.14, that the unit circle is parametrized by the following equations:

$$\cos \theta = \frac{1 - t^2}{1 + t^2}$$

$$\sin \theta = \frac{2t}{1 + t^2}.$$

We can now proceed.

$$
\begin{aligned}
\int \sec \theta \, d\theta &= \int \frac{d\theta}{\cos \theta} \\
&= \int \frac{1 + t^2}{1 - t^2} \cdot \frac{2 \, dt}{1 + t^2} \\
&= \int \frac{2 \, dt}{1 - t^2} \\
&= \int \frac{dt}{1 + t} + \int \frac{dt}{1 - t}
\end{aligned}
$$

[because $2/(1 - t^2) = 1/(1 + t) + 1/(1 - t)$]. Now

$$\int \frac{dt}{1 + t} + \int \frac{dt}{1 - t} = \ln|1 + t| - \ln|1 - t|$$

$$= \ln \left| \frac{1 + t}{1 - t} \right|.$$

The hard part of this problem is finished; we now rewrite the indefinite integral in terms of $\theta$.

$$\frac{1+t}{1-t} = \frac{(1+t)(1+t)}{(1-t)(1+t)}$$

$$= \frac{(1+t)^2}{1-t^2}$$

$$= \frac{1+2t+t^2}{1-t^2}$$

$$= \frac{1+t^2}{1-t^2} + \frac{2t}{1-t^2}$$

$$= \frac{1+t^2}{1-t^2} + \frac{2t/(1+t^2)}{(1-t^2)/(1+t^2)}$$

$$= \frac{1}{\cos\theta} + \frac{\sin\theta}{\cos\theta}$$

$$= \sec\theta + \tan\theta.$$

Therefore,

$$\int \sec\theta \, d\theta = \ln|\sec\theta + \tan\theta|.$$

The substitution $t = \tan(\theta/2) = \sin\theta/(\cos\theta + 1)$ will allow us to integrate not only secant but any rational function of sine and cosine. We introduce some terminology to state the result precisely. An expression $F(x, y)$ of the form

$$a_{00} + a_{10}x + a_{01}y + a_{20}x^2 + a_{11}xy + a_{02}y^2 + \cdots + a_{mn}x^m y^n$$

is called a *polynomial in two variables*; in other words, $F(x, y)$ is a sum of finitely many terms of the form $ax^i y^j$, where $i \geq 0$ and $j \geq 0$ are integers and $a$ is a constant. A *rational function of two variables* $R(x, y)$ is an expression of the form

$$R(x, y) = \frac{F(x, y)}{G(x, y)},$$

where $F(x, y)$ and $G(x, y)$ are polynomials of two variables.

The next theorem says that any rational function of sine and cosine can be integrated to please Archimedes.

**Theorem 2.19.** If $R(x, y)$ is a rational function, then $\int R(\cos\theta, \sin\theta)\, d\theta$ can be integrated explicitly.

*Proof.* The substitution $t = \sin\theta/(\cos\theta + 1)$ of Eq. (7) gives the formulas $\cos\theta = (1 - t^2)/(1 + t^2)$, $\sin\theta = 2t/(1 + t^2)$, and $d\theta = 2dt/(1 + t^2)$. Therefore, after the substitution, the new integrand is a rational function $g(t)$ of the (single) variable $t$ with

$$g(t)dt = g\left(\frac{\sin\theta}{\cos\theta + 1}\right) d\theta = R(\cos\theta, \sin\theta)d\theta.$$

Because every such rational function can be integrated explicitly by partial fractions, there is a function $h(t)$ with $h'(t) = g(t)$. After rewriting $t$ in terms of $\theta$, the method of substitution gives

$$h'\left(\frac{\sin\theta}{\cos\theta + 1}\right) = g\left(\frac{\sin\theta}{\cos\theta + 1}\right) = R(\cos\theta, \sin\theta). \quad \bullet$$

In the example of secant above, $R(x, y) = 1/x$, so that $R(\cos\theta, \sin\theta) = 1/\cos\theta = \sec\theta$. After the substitution, the integrand is $2/(1 - t^2)$, the indefinite integral is $h(t) = \ln|(1 + t)/(1 - t)|$, and

$$\ln\left|\frac{1 + [\sin\theta/(\cos\theta + 1)]}{1 - [\sin\theta/(\cos\theta + 1)]}\right|$$

is the indefinite integral $\int \sec\theta\, d\theta$ (the simplification of the argument to $\ln|\sec\theta + \tan\theta|$ merely makes the formula look nicer).

Here is another illustration of the theorem; let us integrate

$$\int \frac{\sin\theta\, d\theta}{1 - \cos\theta}.$$

Now $R(x, y) = y/(1-x)$; the $\tan(\theta/2)$ substitution gives, after simplifying,

$$\int \frac{\sin\theta \, d\theta}{1-\cos\theta} = \int \frac{2\,dt}{t(1+t^2)}.$$

Partial fractions gives

$$\frac{2}{t(1+t^2)} = \frac{2}{t} - \frac{2t}{1+t^2},$$

and so

$$\int \frac{2\,dt}{t(1+t^2)} = 2\ln|t| - \ln|1+t^2| = \ln\left|\frac{t^2}{1+t^2}\right|.$$

Using Eq. (7), $t = \sin\theta/(1+\cos\theta)$, one can now rewrite the latter expression in terms of $\theta$.

Recall that a ***parametrized curve*** is given by a pair of functions, $\varphi(t)$ and $\psi(t)$. One views such a curve as the path of a bug; at time $t$, the bug is at the point $(x, y) = (\varphi(t), \psi(t))$. We say that a curve can be ***parametrized by rational functions*** if both $\varphi(t)$ and $\psi(t)$ are rational functions.

The key idea behind Theorem 2.19 is that there is a substitution $\theta = u(t)$, namely, $\theta = 2\tan^{-1} t$, and rational functions $\varphi(t)$, $\psi(t)$, and $\delta(t)$ with

$$\cos\theta = \varphi(t), \quad \sin\theta = \psi(t), \quad \text{and } d\theta = \delta(t)\,dt.$$

With this data, $\int R(\cos\theta, \sin\theta)\,d\theta$ changes into $\int g(t)\,dt$, where $g(t)$ is a rational function, and the latter integral can be evaluated explicitly by partial fractions. [It turns out that $\delta(t)$ is necessarily a rational function. On the one hand, $\cos\theta = \varphi(t)$ gives $(d/dt)(\cos\theta) = \varphi'(t)$; on the other hand, the chain rule gives $(d/dt)(\cos\theta) = -\sin\theta(d\theta/dt) = -\sin\theta\,\delta(t) = -\psi(t)\,\delta(t)$. Therefore, $\delta(t) = -\varphi'(t)/\psi(t)$ is a rational function.]

We are going to see that this same idea, but parametrizing other conic sections by rational functions, not only the circle, will allow us to integrate more functions.

***Lemma 2.20.*** The parabola with equation $y^2 = x$, the ellipse with equation $x^2/\alpha^2 + y^2/\beta^2 = 1$, and the hyperbola with equation $x^2/\alpha^2 - y^2/\beta^2 = 1$ can be parametrized by rational functions.

***Remark.*** It can be shown that every conic section can be parametrized by rational functions.

*Proof.* We first consider the parabola. The point $A = (0, 0)$ is on the parabola.

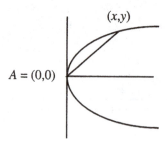

Figure 2.42

The line joining $A$ with another point $(x, y)$ on the parabola has equation $y = tx$ where, of course, $t$ is the slope. The equation of the parabola gives

$$y^2 = t^2x^2 = x$$

so that if $x \neq 0$, we have

$$x = \frac{1}{t^2} = \varphi(t) \quad \text{and} \quad y = \frac{1}{t} = \psi(t).^{14}$$

Next, we consider the ellipse with equation $x^2/\alpha^2 + y^2/\beta^2 = 1$. The point $A = (-\alpha, 0)$ is on the ellipse. The line joining $A$ with another point $(x, y)$ on the ellipse has equation $y = t(\alpha + x)$ where, of course, $t$ is the slope. The equation of the ellipse gives

$$1 - \frac{x^2}{\alpha^2} = \frac{\alpha^2 - x^2}{\alpha^2} = \frac{y^2}{\beta^2} = \frac{t^2(\alpha + x)^2}{\beta^2}.$$

Canceling $\alpha + x$ from both sides gives a linear equation in $x$, and solving for $x$ gives

$$x = \frac{\alpha(\beta^2 - \alpha^2 t^2)}{\beta^2 + \alpha^2 t^2} = \varphi(t).$$

---

[14]Of course, there is a more obvious parametrization of this parabola, namely, $x = t^2$ and $y = t$. The reason for the text's more complicated parametrization is to show that there is one method that parametrizes all conic sections.

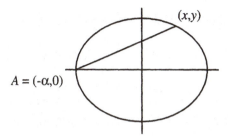

Figure 2.43

Next,

$$y = t(\alpha + x) \;=\; t\left[\alpha + \frac{\alpha(\beta^2 - \alpha^2 t^2)}{\beta^2 + \alpha^2 t^2}\right]$$

$$=\; \frac{2\alpha\beta^2 t}{\beta^2 + \alpha^2 t^2} = \psi(t).$$

We now consider the hyperbola with equation $x^2/\alpha^2 - y^2/\beta^2 = 1$. Let $A = (-\alpha, 0)$, and consider the line joining $A$ with a point $(x, y)$ on the hyperbola. If $t$ is the slope of this line, then $y = t(\alpha + x)$. Replacing $y$ by

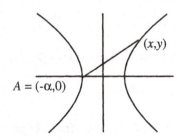

Figure 2.44

$t(\alpha + x)$ in the equation of the hyperbola leads, as for the ellipse, to rational functions. We leave the details to the reader, but we present the formulas:

$$x = \frac{\alpha(\beta^2 + \alpha^2 t^2)}{\beta^2 - \alpha^2 t^2} = \varphi(t)$$

and

$$y = \frac{2\alpha\beta^2 t}{\beta^2 - \alpha^2 t^2} = \psi(t). \quad \bullet$$

***Theorem 2.21.*** If $f(x)$ is a linear or quadratic polynomial and $R(x, y)$ is a rational function, then $\int R(x, \sqrt{f(x)})\,dx$ can be integrated explicitly.

*Proof.* Assume first that $f(x)$ is linear; i.e., $f(x) = ax + b$. The substitution $u = ax + b$ replaces $\sqrt{ax + b}$ by $\sqrt{u}$; we ask the reader to prove that it suffices to prove the theorem in this case. By the lemma, $x = 1/t^2$ and $y = 1/t$. It follows that $R(u, \sqrt{u}) = R(1/t^2, 1/t)$ is a rational function of $t$. As the differential $du = -2\,dt/t^3$, the new integrand $-2R(1/t^2, 1/t)/t^3$ is also a rational function of $t$, and hence it can be integrated explicitly by the method of partial fractions.

Assume that $f(x) = ax^2 + px + q$. Completing the square, which involves the substitution $x = u - p/2a$, replaces $f(x)$ with $au^2 + b$, where $b = q - p^2/4a$; we ask the reader to prove that it suffices to prove the theorem for polynomials of this form.

If $y = \sqrt{f(x)}$, where $f(x) = ax^2 + b$, then $y^2 = ax^2 + b$. There are several possibilities according to the sign of the coefficients: $a = \pm\alpha^2$ and $b = \pm\beta^2$, where $\alpha$ and $\beta$ are positive:

$$\text{(i)}\quad y^2 = \alpha^2 x^2 + \beta^2;$$

$$\text{(ii)}\quad y^2 = \alpha^2 x^2 - \beta^2;$$

$$\text{(iii)}\quad y^2 = -\alpha^2 x^2 + \beta^2$$

($y^2 = -\alpha^2 x^2 - \beta^2$ cannot occur lest the left side be positive and the right side be negative).

Cases (i) and (ii) lead to the equations

$$\frac{y^2}{\beta^2} - \frac{x^2}{(\beta/\alpha)^2} = 1 \quad \text{and} \quad \frac{x^2}{(\beta/\alpha)^2} - \frac{y^2}{\beta^2} = 1,$$

each of whose graphs is a hyperbola. By the lemma, $x = \varphi(t)$ and $y = \psi(t)$ for rational functions $\varphi(t)$ and $\psi(t)$. It follows that $R(x, \sqrt{f(x)}) = R(x, y) = R(\varphi(t), \psi(t))$ is a rational function of $t$. As $\varphi'(t)$ is also a rational function of $t$, so is the new integrand $R(\varphi(t), \psi(t))\varphi'(t)$, and this integral can be integrated explicitly by the method of partial fractions.

Case (iii) leads to the equation

$$\frac{x^2}{(\beta/\alpha)^2} + \frac{y^2}{\beta^2} = 1,$$

the equation of an ellipse. The lemma says that this curve can be parametrized by rational functions, and so the argument can be completed as in the first two cases. ●

Let us illustrate the theorem by evaluating two such integrals. The first involves the square root of a linear polynomial:

$$\int \frac{dx}{1 + \sqrt{x}}.$$

Make the substitution[15] $x = 1/t^2$, so that $dx = -2dt/t^3$. The transformed integral is

$$-2 \int \frac{dt}{t^2(t + 1)}.$$

The algebraic part of partial fractions gives

$$\frac{-2}{t^2(t + 1)} = \frac{2}{t} - \frac{2}{t^2} - \frac{2}{t + 1},$$

so that the indefinite integral is

$$2 \ln|t| - 2 \ln|t + 1| + \frac{2}{t} = 2 \ln\left|\frac{t}{t + 1}\right| + \frac{2}{t}$$

$$= \frac{2}{t} - 2 \ln\left|\frac{t + 1}{t}\right|$$

$$= \frac{2}{t} - 2 \ln\left|1 + \frac{1}{t}\right|.$$

Since $\sqrt{x} = 1/t$, the answer is $2\sqrt{x} - 2\ln(1 + \sqrt{x})$.

---

[15]This is easier using the "obvious" parametrization $x = t^2$ and $y = t$ in the previous footnote.

Here is an illustration involving the square root of a quadratic polynomial.

$$\int \frac{x^3\,dx}{x + \sqrt{4 - x^2}}$$

We will convert this integral into an integral of a rational function, but we will not go through the drudgery of partial fractions to complete the integration. The quadratic $4 - x^2$ is case (iii) of the lemma with $\alpha = 1$ and $\beta = 2$. Thus, as in Lemma 2.20, the parametrization is

$$x = \frac{4 + t^2}{4 - t^2}$$

and

$$y = \sqrt{4 - x^2} = \frac{8t}{4 - t^2}.$$

Differentiation gives

$$dx = \frac{16t\,dt}{(4 - t^2)^2},$$

and so

$$\int \frac{x^3\,dx}{x + \sqrt{4 - x^2}} = \int \frac{16t\,(4 + t^2)^3\,dt}{(4 - t^2)^4(4 + 8t + t^2)}.$$

There are problems when one tries to generalize Theorem 2.21 to square roots of cubic polynomials. For example, can the curve arising from the equation $y^2 = $ cubic be parametrized by rational functions? After all, a line through a point on a cubic curve may intersect the curve in two other points, not just one. Such a question occurs as one begins the study of *elliptic integrals*, that is, integrals that can be put in the form $\int R(x, \sqrt{f(x)})\,dx$, where $f(x)$ is either a cubic or a quartic polynomial. One interesting elliptic integral arises from the integral for the arclength of an ellipse, whence the name (see Exercise 2.57). It can be proved that this integral cannot be integrated explicitly, and so Theorem 2.21 does not extend to cubics. Closely related to elliptic integrals are *elliptic functions*, and the recent solution of Fermat's Last Theorem uses many deep results about elliptic functions in an essential way.

## *Exercises*

**2.52.** Prove the half angle formula for tangent:

$$\tan(\theta/2) = \frac{\sin\theta}{1 + \cos\theta}.$$

(Hint: Use Eq. (2) and Figure 2.41.)

**2.53.** Find the indefinite integral

$$\int \frac{\sin\theta \, d\theta}{2 + \cos\theta}.$$

**2.54.** Find the indefinite integral

$$\int \frac{(\sin\theta - \cos\theta) \, d\theta}{\sin\theta + \cos\theta}.$$

**2.55.** Find the indefinite integral

$$\int \frac{(\sqrt{x} - 1) \, dx}{\sqrt{x} + 1}.$$

**2.56.** Reduce the indefinite integral $\int x^n \sqrt{x^2 + 1} \, dx$, for any integer $n \geq 0$, to an indefinite integral of a rational function.

**2.57.** (i) Show that the ellipse with equation $x^2/a^2 + y^2/b^2 = 1$ can be parametrized by $x = a\cos\theta$ and $y = b\sin\theta$, where $0 \leq \theta < 2\pi$.
(ii) Show that the arclength of this curve is given by the integral

$$2\int_0^\pi \sqrt{a^2 - (a^2 - b^2)\cos^2\theta} \, d\theta.$$

(iii) Show that the $\tan(\theta/2)$ substitution rewrites this as an elliptic integral of the form $\int R(t, \sqrt{f(t)}) \, dt$, where $R(u, v)$ is a rational function of two variables and $f(t)$ is a quartic polynomial.

*Chapter 3*

# Circles and $\pi$

## APPROXIMATIONS

"All right," said the Cat; and this time it vanished quite slowly, beginning
with the end of the tail, and ending with the grin, which remained some
time after the rest of it had gone.

*Lewis Carroll,* Alice in Wonderland

Having seen how to compute the area of triangles, parallelograms,
and polygons, we are now going to consider curved regions. Areas of polygons can be found by first subdividing them into squares or triangles and then
adding up the areas of each of the individual pieces. But not every region
can be so subdivided; for example, a *disk* (i.e., a bounded region in the plane
whose circumference is a circle) cannot be subdivided into (a finite number
of) polygons. The basic idea, now, is to find the exact area of a disk by approximating it by the areas of inscribed polygons. We will give more details
below, but the reader should now realize why the discussion of the area of a
disk will have a different flavor than that of our earlier discussion of areas.

Let us begin by seeing how one might have discovered the formula for
the area of a disk. Let $D$ be a disk with radius $r$ and circumference $c$. Our goal
is to find the usual area formula

$$\text{area}(D) = \tfrac{1}{2}cr$$

(if $c = 2\pi r$, then $\tfrac{1}{2}cr = \pi r^2$).

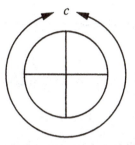

Figure 3.1

Divide $D$ into 4 equal sectors and rearrange them in a row:

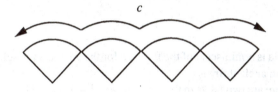

Figure 3.2

Of course, the area remains unchanged, and the total length of the top 4 arcs is still $c$. Now double the area by adding 4 shaded sectors.

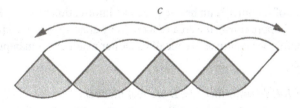

Figure 3.3

If we denote area($D$) by $A$, then the area of this new figure is $2A$. It looks a bit like a parallelogram: each of the scalloped top and bottom edges has length $c$, and each of the two side edges has length $r$ (for they are radii of $D$). Now divide $D$ into more equal sectors, and rearrange them in the same way.

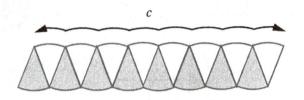

Figure 3.4

The rearrangement, which now looks more like a parallelogram, still has area $2A$, top and bottom of length $c$, and sides of length $r$. If we divide $D$ into a larger number of equal sectors, then the rearranged figure is "almost" a parallelogram with parallel sides of lengths $c$ and $r$. There are two ways to compute the area of this parallelogram. On the one hand, it is $2A$; on the other hand, its area is $cr$. Thus, $2A = cr$, and so

$$A = \tfrac{1}{2}cr.$$

This formula is reminiscent of the formula for the area of a triangle, and it was so viewed in ancient times.

There are two flaws in this preliminary discussion. The circumference $c$ must be computed, and the formula $c = 2\pi r$ is not so easy to prove. But the obvious defect in our preamble is the passage from an "almost" parallelogram to an honest parallelogram. The modern way to deal with approximations, using limits, was introduced by Newton (1642–1727) and Leibniz (1646–1716), independently, in the late seventeenth century; it is the fundamental new idea in calculus. The basic idea of limit is to use approximations to a number $A$ by a sequence of "simpler" numbers to get exact information about $A$. For example, we will approximate the area of a circle by areas of inscribed polygons.

Before we continue, we recall an earlier result about manipulating inequalities.

**Theorem 1.4.** Assume that $b < B$ are real numbers.
(i) If $m$ is positive, then $mb < mB$, whereas if $m$ is negative, then $mb > mB$.
(ii) For any number $N$, positive, negative, or zero, we have

$$N + b < N + B \quad \text{and} \quad N - b > N - B.$$

(iii) Let $c$ and $d$ be positive numbers. If $d < c$, then $1/d > 1/c$, and, conversely, if $1/c < 1/d$, then $c > d$.

Perhaps the first successful study of limits was done by Eudoxus (ca. 400–347 BC), who enunciated a principle, the *Method of Proportions*, that can be found in Euclid's *Elements*. The basic, quite reasonable, assumption made by Eudoxus is that the sequence $\frac{1}{2}, \frac{1}{4}, \ldots, (\frac{1}{2})^n, \ldots$ *has small terms*. Let us discuss this idea in a contemporary way (probably not as Eudoxus might have thought about it). If I say that the sequence $\frac{1}{2}, \frac{1}{4}, \ldots, (\frac{1}{2})^n, \ldots$ has small terms, you might ask me whether it has a term smaller than 0.1. I would reply that if $n = 4$, then $(\frac{1}{2})^4 = \frac{1}{16} = .0625 < 0.1$. You might then ask whether there is a term smaller than .001, and I would say that if $n = 10$, then $(\frac{1}{2})^{10} \approx .00097 < .001$. You might now ask whether there is a term smaller than $10^{-1115}$. If I can always meet your challenge, then you should agree that the sequence does, indeed, have small terms.

***Small Terms Axiom.*** Let $A$ be a positive number. The sequence $\frac{1}{2}A, \frac{1}{4}A, \ldots,$ $(\frac{1}{2})^n A, \ldots$ *has small terms*; that is, for every positive number $E$, there is some integer $\ell$ with $(\frac{1}{2})^\ell A < E$.

The number $E$ is your challenge: Is there a term smaller than $E$? A number $\ell$ with $(\frac{1}{2})^\ell A < E$ is my response to your challenge. The axiom assumes that every challenge $E$ can be met. Of course, my response term $(\frac{1}{2})^\ell A$ depends on the number $E$, as we saw above. We will prove the Small Terms Axiom when we give the modern definition of convergence at the end of this chapter.

Here is my version of the classical Greek notion of approximation (the authentic version is given at the end of this section).

***Definition.*** A sequence $k_1, k_2, k_3, \ldots$ is called *increasing* if $k_1 < k_2 < k_3 < \ldots$; that is, $k_n < k_{n+1}$ for all $n \geq 1$.

A sequence $K_1, K_2, K_3, \ldots$ is called *decreasing* if $K_1 > K_2 > K_3 > \ldots$; that is, $K_n > K_{n+1}$ for all $n \geq 1$.

Of course, there are interesting sequences that are neither increasing nor decreasing, but the sequences arising in the classical proofs of the area and circumference formulas for the disk are of this special type.

***Definition.*** Let $A$ be a positive number. An increasing sequence of positive numbers $k_1 < k_2 < k_3 < \ldots < A$ *approximates $A$ from below*, denoted by $k_* \nearrow A$, if

$$A - k_n < (\tfrac{1}{2})^n A \quad \text{for every } n = 1, 2, 3, \ldots$$

A decreasing sequence $K_1 > K_2 > K_3 > \ldots > A$ *approximates A from above*, denoted by $K_* \searrow A$, if

$$K_n - A < (\tfrac{1}{2})^n K_1 \text{ for every } n = 1, 2, 3, \ldots$$

The next criterion for approximation may be easier to use than the definition.

**Theorem 3.1.** Let $A$ be a positive number.
(i) Let $k_1 < k_2 < k_3 < \ldots < A$ be an increasing sequence. If

$$A - k_1 < \tfrac{1}{2}A$$

and, for every $n \geq 1$,

$$A - k_{n+1} < \tfrac{1}{2}(A - k_n),$$

then $k_* \nearrow A$.
(ii) Let $K_1 > K_2 > K_3 > \ldots > A$ be a decreasing sequence. If

$$K_1 - A < \tfrac{1}{2}K_1$$

and, for every $n \geq 1$,

$$K_{n+1} - A < \tfrac{1}{2}(K_n - A),$$

then $K_* \searrow A$.

*Proof.* (i) It suffices to prove that

$$A - k_n < (\tfrac{1}{2})^n A \text{ for all } n \geq 1.$$

We prove these inequalities by induction on $n \geq 1$. The base step is the given inequality $A - k_1 < \tfrac{1}{2}A$. Let us prove the inductive step. We are assuming that $A - k_{n+1} < \tfrac{1}{2}(A - k_n)$; the inductive hypothesis $A - k_n < (\tfrac{1}{2})^n A$ now gives

$$A - k_{n+1} < \tfrac{1}{2}(A - k_n) < \tfrac{1}{2}\left[(\tfrac{1}{2})^n A\right] = (\tfrac{1}{2})^{n+1} A.$$

(ii) This proof is entirely similar to what was done above, and it is left to the reader.   •

Having checked that a given sequence does, in fact, approximate a number $A$, how can one use this information?

**Theorem 3.2 (Getting Close Principle).** (i) If $B < A$, and if $k_* \nearrow A$, then $B < k_\ell$ for some $\ell$.
(ii) If $B > A$ and if $K_* \searrow A$, then $B > K_m$ for some $m$.

*Remark.* In other words, if $k_* \nearrow A$ and $B < A$, then one can find a term $k_\ell$ between $B$ and $A$; that is, this $k_\ell$ is closer to $A$ than $B$ is.

*Proof.* (i) If, on the contrary, $B \geq k_n$ for all $n \geq 1$, then Theorem 1.4(ii) gives

$$A - B \leq A - k_n \quad \text{for all} \quad n.$$

Now $A - k_n < (\frac{1}{2})^n A$, because $k_* \nearrow A$, so that $A - B \leq A - k_n < (\frac{1}{2})^n A$ for all $n$. Since $A - B$ is positive, we may set $E = A - B$; the Small Terms Axiom says that there is some $\ell$ with $(\frac{1}{2})^\ell A < A - B$. This gives $(\frac{1}{2})^\ell A < A - B < (\frac{1}{2})^\ell A$, a contradiction.
(ii) This proof is similar to that just given.   •

Before continuing, let us consider numbers $C$ and $D$ with decimal expansions $C = 0.c_1 c_2 c_3 \cdots$ and $D = 0.d_1 d_2 d_3 \cdots$, where the digits $c_i$ and $d_i$ lie between 0 and 9. We note that if the first few digits of $C$ and of $D$ coincide, i.e., $c_1 = d_1, c_2 = d_2, c_3 = d_3, \ldots, c_\ell = d_\ell$ and if $c_{\ell+1} < d_{\ell+1}$, then $C < D$. For example, $0.99990 \cdots < 0.99999 \cdots$. After all, $D - C = 0.99999 \cdots - 0.99990 \cdots = 0.00009 \cdots > 0$.
Here is an application of the Getting Close Principle. Let $B$ have an unending string of 9's as its decimal expansion; that is, $B = 0.99999 \cdots$. We have just seen that $B$ is strictly larger than any number $C$ of the form $0.999 \cdots 90$.

**Theorem 3.3.** If $B = .99999 \cdots$, then $B = 1$.

*Proof.* It is plain that $B \leq 1$, so that either $B < 1$ or $B = 1$. Our strategy is to eliminate the first possibility.
We show that the increasing sequence $k_1 = 0.9 < k_2 = 0.99 < k_3 = 0.999 < \ldots < 1$ approximates 1 from below. Note that $k_n = 1 - (\frac{1}{10})^n$ for all $n \geq 1$. Now Exercise 1.4 gives

$$1 - k_n = 1 - [1 - (\tfrac{1}{10})^n] = (\tfrac{1}{10})^n < (\tfrac{1}{2})^n,$$

and we have proved that $k_* \nearrow 1$.

If $B < 1$, then the Getting Close Principle says that there is some $k_\ell$ with $B < k_\ell$; that is, there is some $\ell \geq 1$ so that a number $C$ having only $\ell$ 9's is larger than $B = .99999\cdots$, a never-ending string of 9's. As we remarked above, this is a contradiction. We have eliminated the possibility $B < 1$, and the only remaining option is $B = 1$. •

Is it possible for the same sequence of numbers to approximate two different numbers from below?

**Theorem 3.4.** Let $A$ and $B$ be positive numbers. If either $k_* \nearrow A$ and $k_* \nearrow B$ or if $K_* \searrow A$ and $K_* \searrow B$, then $A = B$.

*Proof.* If $A \neq B$, then one is larger than the other; say, $B < A$. By the Getting Close Principle, there is some $\ell$ with $B < k_\ell$, and this contradicts the hypothesis $k_* \nearrow B$ (which demands that $k_n < B$ for all $n$).

The similar argument that $A = B$ when $K_1, K_2, K_3, \ldots$ approximates both $A$ and $B$ from above is left to the reader. •

**Theorem 3.5 (Sandwich Theorem).** Let $A$ be a positive number, and suppose that $k_* \nearrow A$ and $K_* \searrow A$. If there is a number $c$ for which

$$k_n < c < K_n$$

for all $n$, then $c = A$.

*Proof.* We prove that $c = A$ by eliminating the possibilities $c < A$ and $c > A$. If $c < A$, then the Getting Close Principle says there is some $k_\ell$ with $c < k_\ell$, contrary to the hypothesis $k_n < c$ for all $n$. Similarly, if $c > A$, then the Getting Close Principle says that there is some $K_m$ with $c > K_m$, also contrary to hypothesis. •

*Historical Remarks.* Here is T. L. Heath's baroque translation from classical Greek of Proposition 1 in Book X of Euclid.

> **Method of Proportions (Eudoxus).** Two unequal magnitudes being set out, if from the greater there be subtracted a magnitude greater than its half, and from that which is left a magnitude greater than its half, and if this process is repeated continually, there will be left some magnitude which will be less than the lesser magnitude set out.

This statement is clarified a bit by using symbols, but it still looks rather unwieldy. We make various implicit inequalities explicit.

*Method of Proportions.* Given two positive numbers $B < A$; if there are positive numbers $x_1, x_2, \ldots, x_n, \ldots$ with

$$A > x_1 > \tfrac{1}{2}A,$$

$$A - x_1 > x_2 > \tfrac{1}{2}(A - x_1),$$

$$A - x_1 - x_2 > x_3 > \tfrac{1}{2}(A - x_1 - x_2), \cdots, \text{etc.,}$$

then $A - x_1 - x_2 - \cdots - x_\ell < B$ for some $\ell$.

The Method of Proportions, which combines the notion of approximation and the Getting Close Principle in one statement, was accepted by ancient Greeks as a principle not requiring proof. By invoking the modern idea of small terms, however, we can prove it.

**Theorem (Method of Proportions).** Let $B < A$ be positive numbers. Given numbers $x_1, x_2, \ldots, x_n$ as enunciated by Eudoxus, then

$$A - x_1 - x_2 - \cdots - x_\ell < B$$

for some integer $\ell$.

*Proof.* Define a sequence $k_1, k_2, \ldots, k_n, \ldots$ by

$$k_1 = x_1 \quad \text{and} \quad k_{n+1} = k_n + x_{n+1}.$$

We show first that $k_* \nearrow A$; by Theorem 3.1, it suffices to prove that the increasing sequence $k_1 < k_2 < k_3 < \ldots < A$ satisfies $A - k_1 < \tfrac{1}{2}A$ and $A - k_{n+1} < \tfrac{1}{2}(A - k_n)$ for all $n \geq 1$.

It is given that $A - k_n > x_{n+1}$ for all $n$. Now $A - k_n > 0$, because $x_{n+1}$ is positive, and so $k_n < A$ for all $n$.

By Theorem 1.4(ii), $k_1 = x_1 > \tfrac{1}{2}A$ gives $A - k_1 < A - \tfrac{1}{2}A = \tfrac{1}{2}A$.

Finally, we prove $A - k_{n+1} < \tfrac{1}{2}(A - k_n)$ by induction on $n \geq 1$.

*Base step.* Now $x_2 > \tfrac{1}{2}(A - x_1) = \tfrac{1}{2}(A - k_1)$. After we subtract both sides from $A - k_1$, Theorem 1.4(ii) gives the reverse inequality

$$A - k_2 = (A - k_1) - x_2 < (A - k_1) - \tfrac{1}{2}(A - k_1) = \tfrac{1}{2}(A - k_1).$$

*Inductive step.* Theorem 1.4(ii) and the given inequality

$$x_{n+1} > \tfrac{1}{2}(A - x_1 - \cdots - x_n) = \tfrac{1}{2}(A - k_n)$$

give

$$(A - k_n) - x_{n+1} < (A - k_n) - \tfrac{1}{2}(A - k_n) = \tfrac{1}{2}(A - k_n)$$

for all $n \geq 1$; that is, $A - k_n - x_{n+1} = A - k_{n+1} < \tfrac{1}{2}(A - k_n)$.

We have just proved that $k_* \nearrow A$, so that $A - k_n < (\tfrac{1}{2})^n A$ for all $n \geq 1$.

The Small Terms Axiom provides an integer $\ell$ with $(\tfrac{1}{2})^\ell A < B$. Therefore, $A - k_\ell < (\tfrac{1}{2})^\ell A < B$; that is,

$$A - x_1 - x_2 - \cdots - x_\ell < B. \quad \bullet$$

## Exercises

**3.1.** (i) For $n \geq 1$, define $k_n = 1 - (\tfrac{1}{10})^n$; show that $k_* \nearrow 1$.

(ii) For $n \geq 1$, define $\ell_n = 1 - 2(\tfrac{1}{10})^n$; show that $\ell_* \nearrow 1$.

**3.2.** Show that $1.1, 1.01, 1.001, \ldots$ approximates 1 from above.

**3.3.** Use Theorem 3.1 to prove once again that the sequence $0.9, 0.99, 0.999, \ldots$ approximates 1 from below.

**3.4.** Modify the proof of Theorem 3.3 to show that $\tfrac{1}{3}$ is equal to $.3333\cdots$ (an unending string of 3's).

**3.5** (i) Show that if $M$ is a positive number and $k_* \nearrow A$, then $Mk_* \nearrow MA$.
(ii) Show that if $M$ is a positive number and $K_* \searrow A$, then $MK_* \searrow MA$.
(iii) Use part(i) of this exercise, together with the fact (proved in Theorem 3.3) that the sequence $0.9, 0.99, 0.999, \ldots$ approximates 1 from below, to give a second solution to Exercise 3.4.

**3.6.** Prove that if $a_* \nearrow A$ and $b_* \nearrow B$, then $a_1 + b_1, a_2 + b_2, a_3 + b_3, \ldots$ approximates $A + B$ from below.

**3.7.** Assume that $k_* \nearrow A$.
(i) Show that $k_2, k_3, k_4, \ldots$ approximates $A$ from below.
(ii) For every $m \geq 1$, show that $k_{m+1}, k_{m+2}, k_{m+3}, \ldots$ approximates $A$ from below.

**3.8.** Show that the increasing sequence $k_1 < k_2 < k_3 < \ldots < 1$, where $k_n = 1 - (\tfrac{2}{3})^n$ for all $n \geq 1$, does not approximate 1 from below.

## THE AREA OF A DISK

Who vers'd in geometric lore, would fain
Measure the circle.

*Dante*

We are now going to use the Getting Close Principle to prove that if $D$ and $D'$ are disks with radius $r$ and $r'$, respectively, then $\text{area}(D')/\text{area}(D) = r'^2/r^2$; it will then follow easily that $\text{area}(D) = \pi r^2$. The proof we give is the proof given in Book XII of Euclid's *Elements*. (There are 13 books of Euclid, and they are still being used to teach logical thought after more than two thousand three hundred years. In 1908, C. J. Keyser wrote, "The *Elements* of Euclid is as small a part of mathematics as the *Iliad* is of literature; or as the sculpture of Phidias is of the world's total art.")

The proof of the area formula for a disk will use the fact that areas of inscribed polygons approximate the area of the disk from below (to prove the circumference formula, we will also need the fact that areas of circumscribed polygons approximate the area of the disk from above).

Let $D$ be a disk of radius $r$, and let $P_1$ be a square inscribed in $D$. Bisect all the sides of $P_1$ to get an inscribed regular octagon $P_2$, as in Figure 3.5.

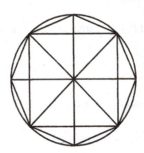

Figure 3.5

Continue this process, so that there is a growing sequence of inscribed regular polygons $P_1$, $P_2$, $P_3$, ... inside of $D$. Note that $P_1$ has 4 sides, $P_2$ has 8 sides, and more generally, $P_n$ is a regular polygon having $2^{n+1}$ sides.

**Theorem 3.6.** The sequence area($P_1$), area($P_2$), area($P_3$), ... approximates $A = $ area($D$) from below: area($P_*$) $\nearrow$ area($D$).

*Proof.* We use the criterion of Theorem 3.1.

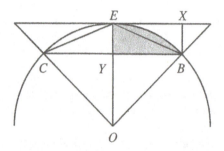

Figure 3.6

It is clear that area($P_n$) < area($D$) for all $n$, for each polygon $P_n$ is inscribed in $D$ and, hence, has smaller area than $D$. Moreover, the sequence of areas is an increasing sequence, for area($P_n$) < area($P_{n+1}$) because $P_n$ is inside of $P_{n+1}$.

If $Q_1$ is a circumscribed square, then Corollary 2.4 gives 2 area($P_1$) = area($Q_1$) > area($D$), so that

$$\text{area}(P_1) = \tfrac{1}{2} \text{area}(Q_1) > \tfrac{1}{2} \text{area}(D),$$

and Theorem 1.4(ii) gives

$$\text{area}(D) - \text{area}(P_1) < \text{area}(D) - \tfrac{1}{2}(D) = \tfrac{1}{2} \text{area}(D).$$

It remains to check the inequalities

$$A - \text{area}(P_{n+1}) < \tfrac{1}{2}[A - \text{area}(P_n)]$$

for all $n \geq 1$. We first describe this inequality geometrically, using Figure 3.6. Let $CB$ be a side of the $2^{n+1}$-gon $P_n$, and let $CE$ and $EB$ be sides of the $2^{n+2}$-gon $P_{n+1}$. On each of the $2^{n+1}$ edges of $P_n$, consider the replica of the region bounded by $CB$ and the arc from $C$ to $B$ through $E$. We focus on $Z_n$, the shaded half of this region, which is bounded by $YB$, $YE$, and the arc from $E$ to $B$. Thus, if $k_n = \text{area}(P_n)$, then there being two copies of $Z_n$ on each of the $2^{n+1}$ edges of $P_n$ gives

$$A - k_n = 2^{n+1}[2 \times \text{area}(Z_n)] = 2^{n+2}\,\text{area}(Z_n).$$

Similarly, if $k_{n+1} = \text{area}(P_{n+1})$, then

$$A - k_{n+1} = 2^{n+2}\,\text{area}(Z_{n+1})$$

($Z_{n+1}$ is bounded by the line $EB$ and the arc from $E$ to $B$). Therefore, to see that $A - k_{n+1} < \frac{1}{2}(A - k_n)$, it suffices to show that $\text{area}(Z_{n+1}) < \frac{1}{2}\,\text{area}(Z_n)$, for then

$$A - k_{n+1} = 2^{n+2}\,\text{area}(Z_{n+1}) < \tfrac{1}{2}2^{n+2}\,\text{area}(Z_n) = \tfrac{1}{2}(A - k_n).$$

Now that the needed inequality has been described geometrically, we prove that it holds.

$$\text{area}(Z_n) < \text{area}(\square BYEX) = 2\,\text{area}(\triangle BYE),$$

so that $\frac{1}{2}\,\text{area}(Z_n) < \text{area}(\triangle BYE)$. Hence,

$$\text{area}(Z_n) = \text{area}(Z_{n+1}) + \text{area}(\triangle BYE)$$

$$> \text{area}(Z_{n+1}) + \tfrac{1}{2}\,\text{area}(Z_n).$$

Subtract $\frac{1}{2}\,\text{area}(Z_n)$ from both sides; Theorem 1.4(ii) [with $N = -\frac{1}{2}\,\text{area}(Z_n)$] gives

$$\tfrac{1}{2}\,\text{area}(Z_n) > \text{area}(Z_{n+1}).$$

The hypotheses of Theorem 3.1 have been verified, and so $\text{area}(P_*) \nearrow \text{area}(D)$.    ●

We now prepare for an application of the Getting Close Principle.

**Lemma 3.7.** Let $D$ and $D'$ be disks with radius $r$ and $r'$, respectively. If $P$ is a polygon inscribed in $D$, and if $P'$ is a similar polygon inscribed in $D'$, then

$$\frac{\text{area}(P')}{\text{area}(P)} = \frac{r'^2}{r^2}.$$

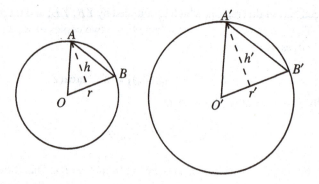

Figure 3.7

*Proof.* Let us first consider a triangle $T$ having one vertex at the center $O$ of the disk $D$ and the other two vertices on the circle, and let $T'$ be a similar triangle in $D'$. Let the height of $T$ from $A$ to the radius $OB$ be denoted by $h$, and let the height of $T'$ from $A'$ to the radius $O'B'$ be denoted by $h'$. Because the triangles $T$ and $T'$ are similar, $h/r = h'/r'$, so that $h' = hr'/r$. Therefore,

$$\frac{\text{area}(T')}{\text{area}(T)} = \frac{\frac{1}{2}h'r'}{\frac{1}{2}hr}$$

$$= \frac{\frac{1}{2}(hr'/r)r'}{\frac{1}{2}hr}$$

$$= \frac{r'^2}{r^2}.$$

We have shown that $\text{area}(T') = (r'^2/r^2)\,\text{area}(T)$.

Now let $\triangle ABC$ be an inscribed triangle in $D$, as in Figures 3.8a and 3.8b. The difference between the two figures is that the center $O$ of the circle is inside $\triangle ABC$ in the first figure, whereas it is outside the triangle in the second. Imagine similar pictures in a disk $D'$ containing an inscribed triangle $\triangle A'B'C'$. In Figure 3.8a, $\triangle ABC$ is dissected into $\triangle OAB$, $\triangle OBC$, and $\triangle OCA$; there is a similar dissection of $\triangle A'B'C'$ in $D'$. We have

$$\frac{\text{area}(\triangle A'B'C')}{\text{area}(\triangle ABC)} = \frac{\text{area}(\triangle O'A'B') + \text{area}\,\triangle(O'B'C') + \text{area}(\triangle O'C'A')}{\text{area}(\triangle OAB) + \text{area}(\triangle OBC) + \text{area}(\triangle OCA)}$$

Because all the triangles in the numerator of the right side have one vertex at

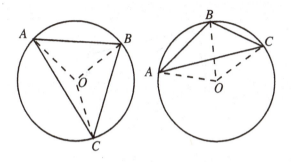

Figure 3.8*a*                    Figure 3.8*b*

$O'$, our first calculation shows that it is equal to

$$\left(\frac{r'^2}{r^2}\right) \text{area}(\triangle OAB) + \left(\frac{r'^2}{r^2}\right) \text{area}(\triangle OBC) + \left(\frac{r'^2}{r^2}\right) \text{area}(\triangle OCA)$$

$$= \left(\frac{r'^2}{r^2}\right) [\text{area}(\triangle OAB) + \text{area}(\triangle OBC) + \text{area}(\triangle OCA)]$$

$$= \left(\frac{r'^2}{r^2}\right) \text{area}(\triangle ABC).$$

Hence, after cancellation, $\text{area}(\triangle A'B'C')/\text{area}(\triangle ABC) = r'^2/r^2$.

Similarly, one can prove that $\text{area}(\triangle A'B'C')/\text{area}(\triangle ABC) = r'^2/r^2$ in Figure 3.8*b* as well, for

$$\text{area}(\triangle ABC) = \text{area}(\triangle OAB) + \text{area}(\triangle OBC) - \text{area}(\triangle OAC).$$

Finally, this same method applies to any inscribed polygon $P$, for $P$ can be dissected into triangles as in Figure 3.9 on the next page.  •

Figure 3.9

We give two proofs of the next theorem, which is the heart of the area formula. The first, more geometric, proof is essentially the one given by Eudoxus; the second, more algebraic, proof is more in the modern spirit.

Here is the ingenious proof of Eudoxus. Notice how it uses the Getting Close Principle to replace the disk by a polygon, thus allowing the last lemma to be used.

**Theorem 3.8.** If $D$ and $D'$ are disks with radius $r$ and $r'$, respectively, then

$$\frac{\text{area}(D')}{\text{area}(D)} = \frac{r'^2}{r^2}.$$

*Proof.* Let us denote area$(D)$ by $A$ and area$(D')$ by $A'$. If, on the contrary, $A'/A \neq r'^2/r^2$, then either $A'/A < r'^2/r^2$ or $A'/A > r'^2/r^2$. In the first case, there is thus some number $M$ with $A'/M = r'^2/r^2$ ($M$ has no obvious geometric interpretation), and $A'/A < A'/M$. Multiplying both sides by $1/A'$ now gives $1/A < 1/M$, and Theorem 1.4(iii) gives $M < A$.

We have seen, in Theorem 3.6, that area$(P_*) \nearrow$ area$(D)$, where $P_n$ is the inscribed regular $2^{n+1}$-gon we constructed. Because $M < A$, the Getting Close Principle says that there is some inscribed polygon $P_\ell$ with $M <$ area$(P_\ell)$. Let $P'_\ell$ be the corresponding polygon in $D'$. By Lemma 3.7,

$$\frac{\text{area}(P'_\ell)}{\text{area}(P_\ell)} = \frac{r'^2}{r^2}.$$

But $A'/M = r'^2/r^2$, so that area$(P'_\ell)/$area$(P_\ell) = A'/M$. Hence

$$\frac{\text{area}(P'_\ell)}{A'} = \frac{\text{area}(P_\ell)}{M}.$$

The left side is smaller than 1 [because $P'_\ell$ is inside of the disk $D'$, hence area$(P'_\ell)$ < area$(D')$ = $A'$], whereas the right side is greater than 1 [for $P_\ell$ was chosen so that $M$ < area$(P_\ell)$], and this is a contradiction.

The other possibility $A'/A > r'^2/r^2$ leads to a contradiction as well (one merely switches the roles of $D$ and $D'$). There is some number $M'$ with $M'/A = r'^2/r^2$, so that $A'/A > M'/A$. Multiplying both sides by $A$ gives $A' > M'$, and one can now repeat the argument above first constructing polygons in $D'$ instead of in $D$.   •

Roughly speaking, the Getting Close Principle allowed us to replace area$(D)$ by area$(P_\ell)$ and area$(D')$ by area$(P'_\ell)$; since area$(P_\ell)$ is "close" to area$(D)$, this is a reasonable thing to do.

Here is a more algebraic proof.

**Theorem 3.9 (= Theorem 3.8).** If $D$ and $D'$ are disks with radius $r$ and $r'$, respectively, then

$$\frac{\text{area}(D')}{\text{area}(D)} = \frac{r'^2}{r^2}.$$

*Proof.* By Theorem 3.6, area$(P'_*) \nearrow$ area$(D')$. Because area$(P'_n)/$ area$(P_n) = r'^2/r^2$ for all $n \geq 1$, by Lemma 3.7, we have area$(P'_n) = (r'^2/r^2)\,$area$(P_n)$ for all $n$. Therefore,

$$\left(\frac{r'^2}{r^2}\right) \text{area}(P_*) \nearrow \text{area}(D').$$

By Theorem 3.6, area$(P_*) \nearrow$ area$(D)$, and so Exercise 3.5(i) gives

$$\left(\frac{r'^2}{r^2}\right) \text{area}(P_*) \nearrow \left(\frac{r'^2}{r^2}\right) \text{area}(D).$$

Theorem 3.4 now shows that area$(D') = (r'^2/r^2)\,$area$(D)$.   •

The familiar definition of $\pi$ as circumference/diameter assumes something about $\pi$ that is not at all obvious, namely, that this ratio is the same for all disks: if $D$ and $D'$ are disks with circumferences $c$ and $c'$ and diameters $d$ and $d'$, why is $c'/d' = c/d$? The definition of $\pi$ given below avoids this question; the area of the unit disk is one specific number.

*Definition*. The number $\pi$ is the area of the unit disk; that is, $\pi$ is the area of a disk with radius 1.

In the next section, we will be able to show that the definition of $\pi$ just given coincides with the familiar definition.

*Corollary 3.10*. If $D$ is a disk with radius $r$, then

$$\text{area}(D) = \pi r^2.$$

*Proof.* If $D'$ is the unit disk, then Theorem 3.8 gives

$$\frac{\text{area}(D')}{\text{area}(D)} = \frac{1^2}{r^2} = \frac{1}{r^2}.$$

Since $\text{area}(D') = \pi$, we have $\text{area}(D) = \pi r^2$, as desired. •

We feel superior to the Greeks because, nowadays, the area formula can be derived routinely using calculus. But let us see whether we have a right to be so smug. The area of a disk of radius $r$ is given by the definite integral

$$A = 2 \int_{-r}^{r} \sqrt{r^2 - x^2} \, dx.$$

Using the substitution: $x = r \sin \theta$, so that $dx = r \cos \theta d\theta$, we see that the indefinite integral is

$$2 \int \sqrt{r^2 - r^2 \sin^2 \theta} \, r \cos \theta d\theta = 2 \int r^2 \cos^2 \theta d\theta;$$

one now uses the double angle formula, $\cos^2 \theta = \frac{1}{2}(1 + \cos 2\theta)$, to obtain

$$\int r^2 (1 + \cos 2\theta) \, d\theta = r^2 (\theta + \frac{1}{2} \sin 2\theta).$$

To evaluate the original definite integral, we must find the new limits of integration: as $x$ varies over the interval $[-r, r]$, $\theta$ varies over the interval $[0, \pi]$. But why is that? For us, $\pi$ means the area of the unit disk; we have not yet established any connection between the area of a circle and its circumference. It follows that it is not yet legitimate to use radian measure, and so it is premature to use limits of integration for the new definite integral that mention $\pi$. Perhaps we should not feel so superior to the Greeks!

### Exercises

**3.9.** A pizzeria charges \$2.50 for a 10″ pizza and \$5.00 for a 15″ pizza (a 10″ pizza is circular with diameter 10 inches). Should four hungry students order four 10″ pizzas or two 15″ pizzas?

**3.10.** Complete the proof of Theorem 3.8 by showing that $A'/A > r'^2/r^2$ leads to a contradiction.

**3.11.** The inscribed polygon $P_n$ consists of $2^{n+1}$ congruent isosceles triangles, each with height $h_n$. Prove that $h_* \nearrow r$; that is, $h_1, h_2, h_3, \ldots$ approximates $r$ from below. (Hint: Using Figure 3.10, express $r - h_{n+1}$ and $r - h_n$ as cosines, and show that $r - h_{n+1} < \frac{1}{2}(r - h_n)$ with the double angle formula.)

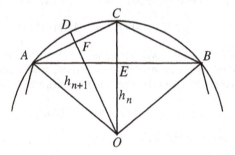

Figure 3.10

**3.12.** For all $n \geq 1$, let $P_n$ be the regular $2^{n+1}$-gon inscribed in a disk $D$ of radius 1 (as constructed in the text). Prove, for all $n \geq 1$, that there are inequalities

$$\text{area}(P_n) < \pi < \text{area}(P_n) + (\tfrac{1}{2})^n \pi.$$

**3.13.** Let $P_n$ be as in Exercise 3.12, and let $s_n$ be the length of a side of $P_n$.
(i) Prove that $s_1 = \sqrt{2}$.
(ii) Prove, for all $n \geq 1$, that

$$s_{n+1} = \sqrt{2 - \sqrt{4 - s_n^2}}.$$

**3.14.** (i) Prove that $\text{area}(P_1) = 2$.
(ii) Prove, for all $n \geq 1$, that

$$\text{area}(P_{n+1}) = 2^n s_n.$$

(Hint: In Figure 3.10, first show that $\text{area}(P_{n+1}) = 2^{n+2} \, \text{area}(\triangle OAC)$; then compute $\text{area}(\triangle OAC)$ by choosing $OC$ as the base and $AE$ as the altitude.)

**3.15.** Use the previous three exercises to estimate $\pi$.
(i) Show that
$$2 < \pi < 4.$$

(ii) Show that
$$2.828427125 < \pi < 3.828427125.$$

(Hint: Use Exercise 3.12 with the estimate $\pi < 4$ of part (i) in the upper bound.)
(iii) Show that $s_2 \approx 0.765366864$, area$(P_3) \approx 3.061147456$, and

$$3.061147456 < \pi < 3.5611467456.$$

(Hint: Use Exercise 3.12 with the estimate $\pi < 4$ from part (i).)
(iv) Show that $s_3 \approx 0.390180643$, area$(P_4) \approx 3.121445144$, and

$$3.121445144 < \pi < 3.371445144.$$

(Hint: Use Exercise 3.12 with the estimate $\pi < 4$ from part (i).)
(v) Repeat this procedure until you can estimate $\pi$ well enough to see that its first digits are 3.14. (Hint: It suffices to show there are inequalities

$$3.140 < \pi < 3.149.)$$

**3.16.** Let $P_n$ be as in Exercise 3.12. Show that

$$\text{area}(P_n) = 2^n \sin(\tfrac{180°}{2^n}).$$

## THE CIRCUMFERENCE OF A DISK

By day and night he measured and calculated; covered enormous quantities of paper with figures, letters, computations, algebraic symbols; his face, which was the face of an apparently sound and vigorous man, wore the morose and visionary stare of a monomaniac; while his conversation, with consistent and fearful monotony, dealt with the proportional number $\pi$ . . .

*Thomas Mann*, The Magic Mountain

We now set the stage for establishing the formula $c = 2\pi r$ for the circumference $c$ of a circle of radius $r$. The proof of this formula was first given by Archimedes (287–212 BC), one of the greatest scientists of antiquity, in the century after Euclid. Like the area formula just proven, it, too, is a formula often quoted but rarely proved in high schools.

Let $Q_1$ denote a circumscribed square, so that its sides are tangent to $D$. Define $Q_2$ to be the circumscribed octagon which is constructed by "cutting off corners" of $Q_1$. In more detail, referring to Figure 3.11, draw the line $\ell$ joining the center $O$ of the disk to a vertex $F$ of $Q_1$; $\ell$ cuts the circle in the point $M$, and the line $HK$, tangent to the circle at $M$, is defined to be one of the sides of $Q_2$; throw away $\triangle FHK$ from $Q_1$. If one repeats this procedure at the other vertices of $Q_1$, one has constructed $Q_2$ by throwing away four triangles (corners) from $Q_1$, one triangle from each vertex of $Q_1$. There are 8 sides: $HK$; $RS$; $TU$; $VW$, and the 4 remnants of the original sides of $Q_1$, namely, $KR$; $ST$; $UV$; $WH$.

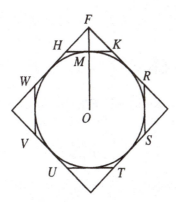

Figure 3.11

This construction can be repeated. Given a circumscribed polygon $Q_n$, which has $2^{n+1}$ sides of equal length, construct $Q_{n+1}$ by connecting $O$ to each of the $2^{n+1}$ vertices of $Q_n$, and then throwing away corners of $Q_n$ in the same way as in the construction of $Q_2$ from $Q_1$. Figure 3.12 shows a portion of $Q_n$ and the smaller $Q_{n+1}$;

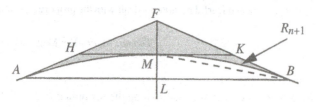

Figure 3.12

$AF$ and $FB$ are remnants of sides of $Q_n$, tangent to $D$ at points $A$ and $B$, respectively. The point $M$ is the midpoint of the arc joining $A$ and $B$, and the tangent $HK$ to the circle at $M$ is one of the sides of $Q_{n+1}$.

**Theorem 3.11.** The sequence area$(Q_1)$, area$(Q_2)$, area$(Q_3)$, ... approximates area$(D)$ from above; that is, area$(Q_*) \searrow$ area$(D)$.

*Proof.* First, it is clear that area$(Q_n) >$ area$(D)$ for all $n \geq 1$ because $D$ is inside of each $Q_n$. Moreover, the sequence of areas is decreasing, area$(Q_n) >$ area$(Q_{n+1})$ for all $n \geq 1$ because $Q_n$ contains $Q_{n+1}$. Second, Corollary 2.4 gives area$(D) >$ area$(P_1) = \frac{1}{2}$ area$(Q_1)$, so that area$(D) > \frac{1}{2}$ area$(Q_1)$. It now follows, from Theorem 1.4(ii), that

$$\text{area}(Q_1) - \text{area}(D) < \text{area}(Q_1) - \tfrac{1}{2}\,\text{area}(Q_1) = \tfrac{1}{2}\,\text{area}(Q_1).$$

Before we check the last criterion in Theorem 3.1, let us describe area$(Q_n) -$ area$(D)$ geometrically. In Figure 3.12, let $R_n$ denote the area of the shaded region bounded by $AF$, $FB$, and the arc joining $AB$. There are $2^{n+1}$ regions congruent to $R_n$, one for each of the vertices of $Q_n$, and

$$\text{area}(Q_n) - \text{area}(D) = 2^{n+1}\,\text{area}(R_n).$$

Of course,

$$\text{area}(Q_{n+1}) - \text{area}(D) = 2^{n+2}\,\text{area}(R_{n+1}),$$

where $R_{n+1}$ is the smaller area bounded by $MK$, $KB$, and the arc joining $M$ and $B$.

Bisect everything with the line $LF$, the perpendicular-bisector of $AB$ (which joins $F$ to the center of $D$), and let $Z_n$ be the right side of $R_n$. Notice that $R_n$ contains two copies of $R_{n+1}$, so that one copy of $R_{n+1}$ lies inside of $Z_n$. Thus,

$$\text{area}(Q_n) - \text{area}(D) = 2^{n+1}\,\text{area}(R_n) = 2^{n+2}\,\text{area}(Z_n),$$

for all $n \geq 1$. It suffices to show that

$$\text{area}(R_{n+1}) < \tfrac{1}{2}\,\text{area}(Z_n),$$

for then

$$\begin{aligned}
\text{area}(Q_{n+1}) - \text{area}(D) &= 2^{n+2}\,\text{area}(R_{n+1}) \\
&< \tfrac{1}{2}2^{n+2}\,\text{area}(Z_n) \\
&= \tfrac{1}{2}[\text{area}(Q_n) - \text{area}(D)].
\end{aligned}$$

Let us return to the proof. We begin by showing that

$$\text{area}(\triangle BKM) < \text{area}(\triangle FKM).$$

Because $LM$ is the altitude of $\triangle BKM$ to the base $KM$,

$$\begin{aligned}
\frac{\text{area}(\triangle BKM)}{\text{area}(\triangle FKM)} &= \frac{\tfrac{1}{2}|LM||KM|}{\tfrac{1}{2}|KM||FM|} = \frac{|LM|}{|FM|} \\
&= \frac{|FL| - |FM|}{|FM|} = \frac{|FL|}{|FM|} - 1.
\end{aligned}$$

By similarity of the triangles $\triangle BFL$ and $\triangle FKM$,

$$\frac{|FL|}{|FM|} - 1 = \frac{|BF|}{|FK|} - 1 = \frac{|BF| - |FK|}{|FK|} = \frac{|BK|}{|FK|}.$$

But the construction of $Q_{n+1}$ from $Q_n$ gives $|BK| = |KM|$. Therefore,

$$\frac{|BK|}{|FK|} = \frac{|KM|}{|FK|} < 1,$$

because the hypotenuse $FK$ of the right triangle $\triangle FKM$ is longer than the leg $KM$. Now

$$\text{area}(R_{n+1}) < \text{area}(\triangle BKM) < \text{area}(\triangle FKM),$$

the last inequality having just been proved. Therefore,

$$2\,\text{area}(R_{n+1}) < \text{area}(R_{n+1}) + \text{area}(\triangle FKM) = \text{area}(Z_n),$$

and so

$$\text{area}(R_{n+1}) < \tfrac{1}{2}\,\text{area}(Z_n). \quad \bullet$$

Let $c$ denote the circumference of $D$, let $p_n$ denote the perimeter of $P_n$, and let $q_n$ denote the perimeter of $Q_n$. In Figure 3.12, it is obvious that

$$p_n < c; \tag{1}$$

after all, $p_n$ is made up of the sides of $P_n$ of the form $AB$, and if $\alpha$ is the arc from $A$ to $B$ through $M$, then $|AB| < \text{length}(\alpha)$, because a straight line is the shortest path between two points. It is not so obvious that $q_n > c$. Why is $|AF| + |FB| > \text{length}(\alpha)$? Archimedes focused on a special property of curves like arcs of circles, namely, the chord joining any two points on a circle lies inside the disk.

**Definition.** Let $\alpha$ be a curve joining a pair of points $A$ and $B$. We say that $\alpha$ is **concave** with respect to $AB$ if every chord $UV$, where $U$ and $V$ are points on $\alpha$, lies inside the region bounded by $\alpha$ and $AB$.

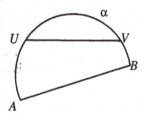

Figure 3.13

Figure 3.13 shows a concave curve, whereas Figure 3.14 gives an example of a curve that is not concave.

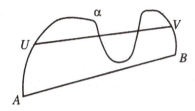

Figure 3.14

**Concavity Principle (Archimedes).** Let $\alpha$ be a curve joining points $A$ and $B$ that is concave with respect to $AB$. If $\beta$ is another concave curve joining $A$ and $B$ that lies inside the region bounded by $\alpha$ and $AB$, then

$$\text{length}(\alpha) > \text{length}(\beta).$$

We are going to use this principle in the special case when $\alpha$ is a path consisting of 2 edges and $\beta$ is an arc of a circle, as in Figure 3.15.

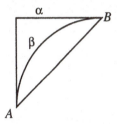

Figure 3.15

As with the Method of Proportions enunciated by Eudoxus, the Concavity Principle of Archimedes was not proved in classical times (it requires an analysis of what one means by the length of a curve, defined nowadays using the arclength formula of calculus). It is remarkable how these ancient thinkers were able to state exactly what was needed to give a coherent proof.

Before proving the perimeter formula, let us show that the Concavity Principle is plausible. Suppose that $\alpha$ is a (necessarily concave) 2-edged path and that the inside path $\beta$ also consists of 2 edges, as in Figure 3.16. We claim that

$$|AC| + |CB| > |AD| + |DB|.$$

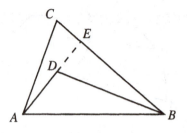

Figure 3.16

This inequality holds, for

$$|AC| + |CB| = |AC| + |CE| + |EB| > |AE| + |EB|$$

(the shortest path from $A$ to $E$ is the straight line $AE$). But

$$|AE| + |EB| = |AD| + |DE| + |EB| > |AD| + |DB|$$

(the shortest path from $D$ to $B$ is the straight line $DB$). We have verified the inequality $|AC| + |CB| > |AD| + |DB|$.

Now let $\beta$ be a 3-edged concave path, as in Figure 3.17.

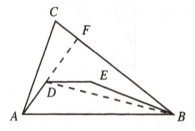

Figure 3.17

Extend $AD$, and let it meet $CB$ in the point $F$. Because the 3-edged path $\beta$ is concave, it is entirely inside the triangle $\triangle AFB$. Let us compute.

$$|AC| + |CB| = |AC| + |CF| + |FB|$$

$$> |AF| + |FB|$$

$$= |AD| + |DF| + |FB|.$$

But $\triangle FDB$, with the 2-edged path $DE$ and $EB$ inside it, is just a replica of Figure 3.16, and so

$$|DF| + |FB| > |DE| + |EB|.$$

Therefore,

$$|AC| + |CB| > |AD| + |DF| + |FB| > |AD| + |DE| + |EB|,$$

as desired.

In Exercise 3.17, this argument is extended to all intermediate $n$-edged concave paths $\beta$ from $A$ to $B$, where $n \geq 2$. It should also be clear, in Figure 3.17, that there are such multi-edged paths with lengths very close to the length of the arc of a circle joining $A$ and $B$; this is the key point in a (modern) proof of the Concavity Principle.

**Lemma 3.12.** If $q_n$ denotes the perimeter of the circumscribed polygon $Q_n$, then $q_n > c$.

*Proof.*

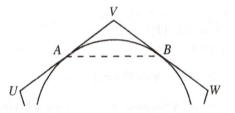

Figure 3.18

As in Figure 3.18, the perimeter of $Q_n$ consists of pairs of edge-pieces ($AV$ is a piece of the edge $UV$, $VB$ is a piece of the edge $VW$), each tangent to the circle from a vertex of $Q_n$. The Concavity Principle says that the length of such paths $AV + VB$ is longer than the arc of the circle joining $A$ and $B$.  •

We are going to give two proofs of the circumference formula; the first is in the classical spirit, the second is in the more modern spirit.

**Theorem 3.13.** Let $D$ be a disk of radius $r$ and circumference $c$. If $\triangle$ is a right triangle with legs of lengths $r$ and $c$, then

$$\text{area}(D) = \text{area}(\triangle).$$

*Proof.* If area$(\triangle) = \frac{1}{2}rc \neq$ area$(D)$, then either $\frac{1}{2}rc >$ area$(D)$ or $\frac{1}{2}rc >$ area$(D)$. We shall reach a contradiction in each of these two cases; equality will then be the only possibility.

Assume first that $\frac{1}{2}rc <$ area$(D)$. Now area$(P_*) \nearrow$ area$(D)$, by Theorem 3.6, so that the Getting Close Principle gives a polygon $P_\ell$ with

$$\tfrac{1}{2}rc < \text{area}(P_\ell).$$

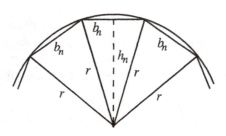

Figure 3.19

Now $P_\ell$ is divided into $2^{\ell+1}$ congruent isosceles triangles, each having height $h_\ell$ and base $b_\ell$. Thus, area$(P_\ell) = 2^{\ell+1}(\frac{1}{2}h_\ell b_\ell)$. The perimeter $p_\ell$ of $P_\ell$ is equal to $2^{\ell+1}b_\ell$, so that

$$\text{area}(P_\ell) = \tfrac{1}{2}h_\ell p_\ell.$$

Because $h_\ell < r$ ($r$ is the hypotenuse of a right triangle having $h_\ell$ as a leg) and $p_\ell < c$, by Eq. (1), we have

$$\text{area}(P_\ell) = \tfrac{1}{2}h_\ell p_\ell < \tfrac{1}{2}rc, \tag{2}$$

and this contradicts the choice of $P_\ell$.

Assume now that $\frac{1}{2}rc >$ area$(D)$. Now area$(Q_*) \searrow$ area$(D)$, by Theorem 3.11, so that the Getting Close Principle gives a polygon $Q_m$ with

$$\tfrac{1}{2}rc > \text{area}(Q_m).$$

Now $Q_m$ is divided into $2^{m+1}$ congruent isosceles triangles, each having height $r$ and base $s_m$, say. Thus, area$(Q_m) = 2^{m+1}(\frac{1}{2}rs_m)$. The perimeter $q_m$ of $Q_m$ is equal to $2^{m+1}s_m$, so that

$$\text{area}(Q_m) = \tfrac{1}{2}rq_m.$$

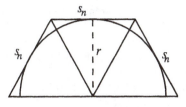

Figure 3.20

By Lemma 3.12, $q_m > c$. It follows that

$$\text{area}(Q_m) = \tfrac{1}{2}rq_m > \tfrac{1}{2}rc, \tag{3}$$

and this contradicts the choice of $Q_m$.

It follows that $\tfrac{1}{2}rc = \text{area}(D)$, as desired.  •

Thus, the circumscribed polygons $Q_n$ are needed to eliminate the possibility $\tfrac{1}{2}rc > \text{area}(D)$.

Perhaps this theorem is related to the classical problem of "squaring the circle." Given a disk $D$, we have just found a right triangle whose area is that of $D$. Can a square having area $D$ be constructed with compass and straight-edge?

Here is a second proof of the circumference formula.

**Theorem 3.14 (= Theorem 3.13).** *If $D$ is a disk of radius $r$ and circumference $c$, then*

$$c = 2\pi r.$$

*Proof.* Inequality (2) in the preceding proof gives $\text{area}(P_n) < \tfrac{1}{2}rc$, and inequality (3) gives $\tfrac{1}{2}rc < \text{area}(Q_n)$. Hence, for all $n \geq 1$, we have

$$\text{area}(P_n) < \tfrac{1}{2}rc \nleqq \text{area}(Q_n).$$

Now $\text{area}(P_*) \nearrow \text{area}(D)$, by Theorem 3.6, and $\text{area}(Q_*) \searrow \text{area}(D)$, by Theorem 3.11. It follows from the Sandwich Theorem, Theorem 3.5, that $\tfrac{1}{2}rc = \text{area}(D) = \pi r^2$, and so $c = 2\pi r$.  •

Both proofs of the circumference formula require the same amount of work; the second proof uses the Sandwich Theorem in place of the Getting Close Principle.

*Corollary 3.15.* If $c$ is the circumference of a circle having diameter $d$, then $\pi = c/d$.

*Proof.* Since $c = 2\pi r$ and $d = 2r$, we have $c/d = \pi$.    •

Only now, having determined that the circumference of the unit disk is $2\pi$, can one use radian measure.

The perimeter of the unit disk being $2\pi$ is the reason $\pi$ is so denoted, for it is the first letter of the Greek word meaning *perimeter*. The symbol $\pi$ was introduced by William Jones in 1706; some earlier notations were $\pi/\delta$ (Oughtred, 1652), $e$ (Sturm, 1689), and $c/r$ (for $2\pi$) (De Moivre, 1698).

Let us compare the classical determination of the circumference of a circle of radius $r$ with the standard calculation nowadays done in calculus. First of all, one develops the arclength formula (the idea behind which is plainly visible in the work of Archimedes, for arclength is defined as a limit of lengths of polygonal paths). If a curve $\gamma$ has a parametrization:

$$x = f(t), y = g(t), \text{ where } a \leq t \leq b,$$

then the length $L$ of $\gamma$ is given as

$$L = \int_a^b \sqrt{f'(t)^2 + g'(t)^2}\, dt.$$

Parametrize the circle by $x = f(t) = r \cos t$ and $y = g(t) = r \sin t$, where $0 \leq t \leq 2\pi$; now $f'(t)^2 + g'(t)^2 = r^2(\sin^2 t + \cos^2 t) = r^2$, and $L = 2\pi r$. However, this is more subtle than it first appears, for one must ask how the number $\pi$, defined as the area of the unit circle, enters into the parametrization. The answer, of course, is via radian measure, which presupposes the perimeter formula (it is a bad pun to call this circular reasoning).

In the next chapter, we shall prove that $\pi$ is an irrational number ($\pi$ is even known to be a *transcendental number*; that is, $\pi$ is not a root of any polynomial whose coefficients are rational numbers). The approximation $22/7 = 3.1428571 \cdots$ to $\pi$ has been used for over two thousand years. Archimedes calculated $\pi$ by inscribing and circumscribing regular polygons in and around a circle of radius 1. Using a regular polygon with 96 sides (he began with a regular hexagon and then doubled it four times), he obtained

$$3.1408 \approx 3 + \tfrac{10}{71} < \pi < 3 + \tfrac{1}{7} \approx 3.1428.$$

In Exercises 3.15 and 3.16, we showed ways of estimating $\pi$ using inscribed polygons. Here are the first 1001 decimal places of $\pi$ obtained by pressing a computer key:

$$\pi = 3.1415926535897932384626433832795028841971693$$
$$9937510582097494459230781640628620899862803482534$$
$$2117067982148086513282306647093844609550582231725$$
$$3594081284811174502841027019385211055596446229489$$
$$5493038196442881097566593344612847564823378678316$$
$$5271201909145648566923460348610454326648213393607$$
$$2602491412737245870066063155881748815209209628292$$
$$5409171536436789259036001133053054882046652138414$$
$$6951941511609433057270365759591953092186117381932$$
$$6117931051185480744623799627495673518857527248912$$
$$2793818301194912983367336244065664308602139494639$$
$$5224737190702179609437027705392171762931767 52384$$
$$6748184676694051320005681271452635608277857713427$$
$$5778960917363717872146844090122495343014654958537$$
$$1050792279689258923542019956112129021960864034418$$
$$1598136297747713099605187072113499999983729780499$$
$$5105973173281609631859502445945534690830264252230$$
$$8253344685035261931188171010003137838752886587533$$
$$2083814206171776691473035982534904287554687311595$$
$$6286388235378759375195778185778053217122680661300$$
$$19278766111959909216420198 93 \cdots$$

In the nineteenth century, decimal places of $\pi$ were calculated by hand, using variants of the power series expansion of arctan. In 1840, Johann Dase computed the first 205 decimal places of $\pi$; in 1873, William Shanks published his calculation of $\pi$ to 707 decimal places in the Proceedings of the Royal Society, London (he also calculated several hundred decimal places of $e$). Nowadays, computers evaluate $\pi$ using certain iterative algorithms and not with inscribed polygons or power series. To hone computing skills and to test certain error estimates, Y. Kanada has computed over 6 billion (i.e., $6.44 \times 10^9$) decimal places of $\pi$.

In the Bible, I Kings 7:23, there is a description of a circular pool in the palace of King Solomon.

וַיַּעַשׂ אֶת הַיָּם מוּצָק עֶשֶׂר בָּאַמָּה מִשְּׂפָתוֹ

עַד שְׂפָתוֹ עָגֹל סָבִיב וְחָמֵשׁ בָּאַמָּה קוֹמָתוֹ

וְקָוֶה שְׁלֹשִׁים בָּאַמָּה יָסֹב אֹתוֹ סָבִיב

Here is the King James translation.

> And he made a molten sea, ten cubits from one brim
> to the other;
> it was round all about, and his height was five cubits;
> and a line of thirty cubits did compass it round about.

Some have concluded that this is a biblical proof that $\pi = 3$ (one does not expect decimal expansions in the Bible, but circumference 31 is more accurate). On the other hand, one can use the original Hebrew text of this verse to get a better value for $\pi$. For centuries, the 22 letters of the Hebrew alphabet have been assigned numerical values as follows:

| | | | | | | | | |
|---|---|---|---|---|---|---|---|---|
| א | 1 | aleph | ח | 8 | kheth | ע | 70 | ayin |
| ב | 2 | beth | ט | 9 | teth | פ | 80 | pe |
| ג | 3 | gimel | י | 10 | yod | צ | 90 | tsade |
| ד | 4 | daled | כ | 20 | kaph | ק | 100 | koph |
| ה | 5 | he | ל | 30 | lamed | ר | 200 | resh |
| ו | 6 | vav | מ | 40 | mem | ש | 300 | shin |
| ז | 7 | zayin | נ | 50 | nun | ת | 400 | tav |
| | | | ס | 60 | samakh | | | |

In the mystic Qabala of the Middle Ages, one assigned a number to each Hebrew word, namely, the sum of the values of each of its letters. Let us return to the Biblical verse. The Hebrew word kav קוה is used for "circumference" (see the bottom line; the initial letter vav ו in וקוה—Hebrew is read from right to left— means "and"). However, the usual spelling of kav is קו without the final letter he ה as in the text. The value Qabala assigns to קו is 106, whereas the value it assigns to קוה is 111. Now

$$\frac{30}{10} \times \frac{111}{106} = 3.141509$$

is an approximation to $\pi$ that is good to 4 decimal places; that is, it is accurate to one part in ten thousand. (We do not insist that you accept Qabalistic methods.)

King Solomon's pool ("sea") was 3-dimensional and, several verses later, the Biblical text says that the thickness of the wall was about the thickness of a hand. In the second century, Rabbi Nehemiah wrote that if the thickness was about $\frac{1}{14}$ of a cubit, then using the approximation of $\frac{22}{7}$ for $\pi$ gives a reasonable measure of the area if the diameter is the outside diameter whereas the circumference is the inside circumference. Perhaps it is simplest, however, if one just says that King Solomon's pool was circular, but not a perfect circle!

In 1897, the Indiana State House of Representatives unanimously passed House Bill 246; here is the bill in its entirety.

House Bill No. 246

A bill for an act introducing a new mathematical truth and offered as a contribution to education to be used only by the State of Indiana free of cost by paying any royalties whatever on the same, provided it is accepted and adopted by the official action of the legislature of 1897.

*Section 1.* Be it enacted by the General Assembly of the State of Indiana: It has been found that a circular area is to the square on a line equal to the quadrant of the circumference as the area of an equilateral rectangle is to the square on one side. The diameter employed as the linear unit according to the present rule in computing the circle's area is entirely wrong, as it represents the circle's area one and one-fifth times the area of a square whose perimeter is equal to the circumference of the circle. This is because one-fifth of the diameter fails to be represented four times in the circle's circumference. For example: if we multiply the perimeter of a square by one-fourth of any line one-fifth greater than one side, we can in like manner make the square's area to appear one fifth greater than the fact, as is done by taking the diameter for the linear unit instead of the quadrant of the circle's circumference.

*Section 2.* It is impossible to compute the area of a circle on the diameter as the linear unit without trespassing upon the area outside of the circle to the extent of including one- fifth more area than is contained within the circle's circumference, because the square on the diameter produces the side of a square which equals nine when the arc of ninety degrees equals eight. But taking the quadrant of the circle's circumference for the linear unit, we fulfill the requirements of both quadrature and rectification of the circle's circumference. Furthermore, it has revealed the ratio of the chord and arc of ninety degrees, which is as seven to eight, and also the ratio of the diagonal and one side of a square which is as ten to seven, disclosing the fourth important fact, that the ratio of the diameter and circumference is as five-fourths to four; and because of these facts and the further fact that the rule in present use fails to work both ways mathematically, it should be discarded as wholly wanting and misleading in its practical applications.

*Section 3.* In further proof of the value of the author's proposed contribution to education, and offered as a gift to the State of Indiana, is the fact of his solutions of the trisection of the angle, duplication of the cube and quadrature of the circle having been already accepted as contributions to science by the American Mathematical Monthly, the leading exponent of mathematical thought in this country. And be it remembered that these noted problems had been long since given up by scientific bodies as unsolvable mysteries and above man's ability to comprehend.

I display the bill for two reasons. The first, of course, is that one ought to see what one is criticizing. Second, we have here a prime example of how not to write mathematics. I will try to decipher the beginning of Section 1, but it is so poorly written that I can only guess at its meaning, if any. Given a circle of radius $r$, its "circular area" is $\pi r^2$; "quadrant" means one-fourth, so that a "quadrant of a circumference" seems to mean $2\pi r/4 = \pi r/2$; "square" seems to mean the algebraic operation of multiplying a number by itself, whereas "equilateral rectangle" seems to mean the geometric figure we call a square (otherwise why not call it a square?). Thus, the text of the bill seems to say that $\pi r^2/(\pi r/2)^2 = r^2/r^2$; that is, the bill wants to legislate the value of $\pi$ to be 4 (it appears there are at least three other values for $\pi$ elsewhere in the first two sections, none of which seem correct), a much worse approximation than that of Archimedes, not to mention that of William Shanks, and not even as good as that occurring in the Bible without the Qabala.

The remainder of this account is taken from the article by W. E. Edington that appeared in the Proceedings of the Indiana Academy of Science in 1935.

> The author of the bill was Edwin J. Goodwin, MD, of Solitude, Posey County, Indiana, and the bill was introduced by his representative Mr. Taylor I. Record. $\cdots$ Following the introduction of the bill, it was referred to the House Committee on Canals. Just why it should be referred to this committee, frequently called the Committee on Swamp Lands, is difficult to understand. $\cdots$

On January 20, 1897, the following item appeared in The Indianapolis Sentinel:

### To Square the Circle

#### Claims Made That This Old Problem Has Been Solved.

The bill telling how to square a circle, introduced in the House by Mr. Record, is not intended to be a hoax. Mr. Record knows nothing of the bill with the exception that he introduced it by request of Dr. Edwin Goodwin of Posey County, who is the author of the demonstration. The latter and State Superintendent of Public Instruction Geeting believe that it is the long-sought solution of the problem, and they are seeking to have it adopted by the legislature. Dr. Goodwin, the author, is a mathematician of note. He has it copyrighted and his proposition is that if the legislature will indorse the solution he will allow the state to use the demonstration in its textbooks free of charge. The author is lobbying for the bill.

The next day, the Indianapolis Sentinel had the following report.

### Mathematical Bill Passed

Record's bill containing the discovery of Dr. Goodwin, of Posey County, for computing the area of a circle, was handed down upon second reading. Mr. Nicholson (chairman of the Committee on Education) explained that Dr. Goodwin had a copyright on his discovery and had offered this bill in order that it might be free to the schools of Indiana. The bill was taken up and passed under suspension of rules. This is the strangest bill that has ever passed an Indiana Assembly. It reads as follows: ...

By chance, Professor C. A. Waldo, a mathematics professor from Purdue University, happened to hear of the proposed law, and he convinced the State Senators to send the bill back to committee, from which it has never emerged. The following account appeared in the Indianapolis News on February 13, 1897.

### The Mathematical Bill

#### Fun Making in the Senate Yesterday Afternoon

Representative Record's mathematical bill legalizing a formula for squaring the circle was brought up and made fun of. The Senators made bad puns about it, ridiculed it and laughed over it. The fun lasted half an hour. Senator Hubbell said that it was not meet for the Senate, which was costing the State $250 a day, to waste its time in such frivolity. He said that in reading the leading newspapers of Chicago and the East, he found that the Indiana State Legislature had laid itself open to ridicule by the action already taken on the bill. He thought consideration of such a proposition was not dignified or worthy of the Senate. He moved the indefinite postponement of the bill, and the motion carried.

I must comment on Section 3 of the bill. One of the early triumphs of modern algebra was the proof that the three classical problems stated therein are impossible to solve *given the ground rules of the problems*: the only permissible tools are a compass and a straight-edge: given two points $P$ and $Q$, the compass can only draw the circles with radius $|PQ|$ and center either $P$ or $Q$; the straight-edge can only draw the line through $P$ and $Q$. In 1837, sixty years before House Bill 246, P. L. Wantzel proved that one can neither trisect a 60° angle nor duplicate the cube using compass and straight-edge in the permissible way. In 1882, fifteen years before the bill, F. Lindemann proved that one cannot square the circle (it is a consequence of his theorem that $\pi$ is transcendental). Proofs of these assertions can be found in many places; for example, see [Rotman, Chapter 4] and [Hadlock].

The State of Indiana did not pass a law specifying a value of $\pi$, although they did come close to doing so. The following letter to the Editor appeared in *The Mathematical Intelligencer*, vol. 7, No. 4, p.6 (1985).

> I enjoyed David Singmaster's article "The Legal Values of Pi" in *The Mathematical Intelligencer*, Vol 7, No. 2, pp. 60-72 (1985). While House Bill No. 246 may have suffered a dilatory fate, a German bill to rescue $\pi$ from irrationality fared better. In fact, such a bill was passed (I suspect sometime after the invention of the automobile) in which the legal value of $\pi$ was set equal to 3.12 (precisely) for purposes of computing the annual tax on automobiles registered in Germany.
>
> How can taxes and $\pi$ possibly be related? Well, the vehicular tax around here is a linear function of engine displacement. And although the displacement of a car that I had bought in 1983 was 1832.248 cm$^3$, my tax bill was based on 1820 cm$^3$. With a piston travel of 9 cm and a bore of 8.05 cm, the implied value of $\pi$ for a 4 cylinder engine is 3.120592913... – rather considerably off the true value, especially considering local preference for precision.
>
> A little later I bought a similar car with the same piston travel but an

increased bore of 8.4 cm. Since I was billed for a displacement of only 1981 cm$^3$, the implied value of $\pi$ has now dropped 3.119488536... Does the German legal value of $\pi$ change with the seasons, or perhaps decrease monotonically with time (as some physicists have hypothesized for the gravitational constant) – or *what* is going on?

My guess is that the exact legal value of $\pi$ for purposes of the German car tax remains fixed at 3.12. The extra digits in the above two values simply result from an additional rule: rounding the displacement volume to the nearest integer (in cm$^3$).

But why did the fiscal authorities pick 3.12 for $\pi$? The value 3.14 would have been much closer and the value 3.15 would have even gained the tax people more good marks. If one of the readers knows the answer, I would like to know.

*Manfred R. Schroeder Drittes Physikalisches Institut Universität Göttingen Federal Republic of Germany*

## *Exercises*

**3.17.** Let $\triangle ABC$ be a triangle in the plane, as in Figure 3.17, and let $\beta$ be a concave polygonal path from $A$ to $B$ lying wholly inside it. If $\beta$ has $n \geq 2$ edges, prove that $|AC| + |CB| > \text{length}(\beta)$. (This exercise is to make the Concavity Principle more plausible, and so this principle may not be used in the proof.)

**3.18.** Show, in Figure 3.21, that $|AC| + |CD| + |DB| > |AE| + |EB|$.

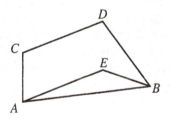

Figure 3.21

**3.19.** True or false: the arclength of the curve $y = x^2$ between 0 and 1 is greater than the arclength of the curve $y = x^3$ between 0 and 1. It is not fair to integrate.

**3.20.** The front wheel of a tricycle is 3 feet in diameter, while its two rear wheels are 2 feet in diameter. If, on a straight road, the front wheel makes 64 revolutions, how many revolutions do the rear wheels make?

## SEQUENCES

The old order changeth, yielding place to new.
*Alfred Lord Tennyson,* Idylls of the King

We begin this section by showing that the Greek notion of approximation is inadequate; we then give the modern definition of convergence and show that it solves the problems of the old. Further study of limits is left to a course in real variables.

Exercise 3.8 signals a problem with the Greek notion of approximation: the sequence $k_n = 1 - (\frac{2}{3})^n$ does not approximate 1 from below. The first few terms of this sequence are:

$$.333, .556, .708, .803, .868, .912, .942, .961, .974, .983, .988, \ldots.$$

The terms get as close to 1 as one wishes, so that this sequence ought to approximate 1 from below, but it does not: $1 - k_n = (\frac{2}{3})^n > (\frac{1}{2})^n$, by Exercise 1.4, and we do not have $1 - k_n < (\frac{1}{2})^n$ as required.

A worse example is provided by the sequence with $n$th term $K_n = \frac{1}{n}$. This sequence $1, \frac{1}{2}, \frac{1}{3}, \frac{1}{4}, \ldots$ does not approximate $A = 0$ from above, for the definition of approximation demands that $\frac{1}{n} < (\frac{1}{2})^n$ for all $n \geq 1 (K_1 = 1$ here). But Theorem 1.5 gives $2^n > n$ for all $n \geq 1$, and now Theorem 1.4(iii) gives $(\frac{1}{2})^n < \frac{1}{n}$ for all $n \geq 1$, an inequality in the wrong direction.

The problem lies in our insisting that terms of a sequence get small at about the same rate as do the powers of $\frac{1}{2}$, and this is too restrictive. It is important that the terms get small, but it is less important that they get small at some particular rate.

There are also algebraic problems with the classical notion. Two useful properties do hold. In Exercise 3.5, we saw that if $a_* \nearrow A$, then $Ma_* \nearrow MA$ for any $M > 0$; in Exercise 3.6, we saw that if $a_* \nearrow A$ and $b_* \nearrow B$, then $a_* + b_* \nearrow A + B$. We now ask whether $a_* \nearrow A$ and $b_* \nearrow B$ imply $a_* b_* \nearrow AB$; that is, does the sequence $a_1 b_1, a_2 b_2, a_3 b_3, \ldots$ approximate $AB$ from below? Of course, $a_n < A$ and $b_n < B$ imply $a_n b_n < AB$; moreover, $a_n < a_{n+1}$ implies $a_n b_n < a_{n+1} b_n$, and $b_n < b_{n+1}$ implies $a_{n+1} b_n < a_{n+1} b_{n+1}$, so that the sequence $\{a_n b_n\}$ is increasing. The real question is whether or not the inequality $AB - a_n b_n < (\frac{1}{2})^n AB$ holds for all $n$. The standard "trick" is to add and subtract the same number:

$$AB - a_n b_n = AB - Ab_n + Ab_n - a_n b_n$$

$$= A(B - b_n) + (A - a_n)b_n$$

$$< (\tfrac{1}{2})^n AB + (\tfrac{1}{2})^n Ab_n$$

$$< (\tfrac{1}{2})^n AB + (\tfrac{1}{2})^n AB \qquad [\text{ for } b_n < B]$$

$$= 2 \times (\tfrac{1}{2})^n AB = (\tfrac{1}{2})^{n-1} AB.$$

Thus, $AB - a_n b_n < (\tfrac{1}{2})^{n-1} AB$. However, we need $AB - a_n b_n < (\tfrac{1}{2})^n AB$; hence, $AB - a_n b_n$ is not small enough, for $(\tfrac{1}{2})^n AB < (\tfrac{1}{2})^{n-1} AB$. We have shown only that the obvious way to prove that the sequence $a_1 b_1$, $a_2 b_2, a_3 b_3, \ldots$ approximates $AB$ from below does not seem to work, but we have not proved that the result is false; perhaps one can prove it with some other approach. To show that the statement really is false, however, we must present a concrete counterexample. If $k_n = 1 - (\tfrac{2}{5})^n$, then $k_* \nearrow 1$, because

$$1 - [1 - (\tfrac{2}{5})^n] = (\tfrac{2}{5})^n < (\tfrac{1}{2})^n$$

(using Exercise 1.4, for $\tfrac{2}{5} = 0.4 < 0.5 = \tfrac{1}{2}$). However, we now show that the sequence of squares $k_1 k_1, k_2 k_2, k_3 k_3, \ldots$ does not approximate 1 from below:

$$1 - (k_n)^2 = 1 - [1 - (\tfrac{2}{5})^n]^2 = 2(\tfrac{2}{5})^n - (\tfrac{2}{5})^{2n}.$$

To satisfy the definition, this last number must be less than $(\tfrac{1}{2})^n$ for all $n \geq 1$. But if $n = 1$,

$$2(\tfrac{2}{5})^n - (\tfrac{2}{5})^{2n} = \tfrac{4}{5} - \tfrac{4}{25} = \tfrac{16}{25} = 0.64 > \tfrac{1}{2};$$

for good measure, if $n = 2$, then

$$2(\tfrac{2}{5})^n - (\tfrac{2}{5})^{2n} = \tfrac{8}{25} - \tfrac{16}{625} = \tfrac{184}{625} = 0.2944 > \tfrac{1}{4}.$$

(To disprove the statement "$A - k_n < (\tfrac{1}{2})^n A$ for all $n \geq 1$," it would have been enough to find just one $n$ for which it is false; one fly in the ointment suffices.) It should now be clear that the Greek notion of approximation, although good enough for $\pi$, is not good enough for other purposes.

The modern definition of convergence of sequences will drop the classical assumption that all numbers involved are positive. There is a modest price we have to pay for this added generality: *absolute value* must be introduced.

*Definition*. The **absolute value** $|a|$ of a number $a$ is defined as follows: if $a \geq 0$, then $|a| = a$; if $a < 0$, then $|a| = -a$.

Thus, $|a| = \pm a$ for all real numbers $a$; it follows that $|a|^2 = a^2$ for all $a$ (we will prove that $(-1) \times (-1) = +1$ in the next chapter).

Here are some basic properties of absolute value.

**Lemma 3.16.** Let $x$ and $y$ be numbers.
(i) $|x| \geq 0$, and $|x| = 0$ if and only if $x = 0$.
(ii) $|-x| = |x|$.
(iii) $|x + y| \leq |x| + |y|$.
(iv) $|xy| = |x||y|$.

*Proof.* (i) If $x \geq 0$, then $|x| = x \geq 0$; if $x < 0$, then $-x > 0$, and so $|x| = -x > 0$.

Of course, $|0| = 0$. Conversely, if $|x| = 0$, then $\pm x = 0$, for $|x| = \pm x$, and so $x = 0$.

(ii) If $x \geq 0$, then $|x| = x$ and $|-x| = -(-x) = x$. If $x < 0$, then $|x| = -x$; on the other hand, $|-x| = -x$ because $-x > 0$.

(iii) We show first that if $p$ and $q$ are positive numbers with $p^2 \leq q^2$, then $p \leq q$. Now $q^2 - p^2 = (q + p)(q - p)$ is positive, by hypothesis. As both $q$ and $p$ are positive, $q + p$ is positive. It follows that $q - p \geq 0$ [otherwise $(q + p)(q - p)$ would be negative]. We conclude that $q - p \geq 0$; that is, $p \leq q$.

Now $|x + y|^2 = (x + y)^2 = x^2 + 2xy + y^2$, whereas $(|x| + |y|)^2 = |x|^2 + 2|x||y| + |y|^2$. But $2xy \leq 2|x||y|$, so that $|x + y|^2 \leq (|x| + |y|)^2$ and, hence, the result just proved gives $|x + y| \leq |x| + |y|$.

(iv) Because $|a|^2 = a^2$ for all numbers $a$, we look at the squares.

$$|xy|^2 = (xy)^2 = x^2y^2 = |x|^2|y|^2 = (|x||y|)^2.$$

Taking square roots, $|xy| = \pm|x||y|$; as both $|xy|$ and $|x||y|$ are nonnegative, however, the desired equality holds. •

Absolute value will enable us to speak of the *distance* between two numbers (regarding them as points on the $x$-axis). Denote the distance between points $a$ and $b$ by $d(a, b)$. Now any notion of distance should, at the very least, satisfy the following three properties.

(1) $d(a, b) \geq 0$ for all $a$ and $b$, and $d(a, b) = 0$ if and only if $a = b$.

The distance between points is a nonnegative real number (two points cannot be $-2$ units apart); moreover, distinct points have some positive distance between them.

(2) $d(a, b) = d(b, a)$ for all $a$ and $b$.

The distance from here to there is the same as the distance from there to here.

(3) **Triangle inequality**: $d(a, b) \leq d(a, c) + d(c, b)$ for all $a$, $b$ and $c$.

The distance $d(a, b)$ measures the shortest path between $a$ and $b$; any path through a third point $c$ is at least as long.

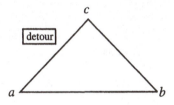

Figure 3.22

**Definition.** If $a$ and $b$ are real numbers, the **distance** from $a$ to $b$ is $|a - b|$.

The next theorem shows that $d(a, b) = |a - b|$ satisfies the necessary three properties a distance must have.

**Theorem 3.17.** Let $a$, $b$, and $c$ be numbers.
 (i) $|a - b| \geq 0$ and $|a - b| = 0$ if and only if $a = b$.
 (ii) $|a - b| = |b - a|$.
 (iii) $|a - b| \leq |a - c| + |c - b|$.

*Proof.* (i) Set $x = a - b$ in Lemma 3.16(i).
 (ii) Because $b - a = -(a - b)$, this follows from Lemma 3.16(ii).
 (iii). Apply Lemma 3.16(iii) to $a - b = (a - c) + (c - b)$.    •

For any real number $x$, we may interpret $|x| = |x - 0|$ as the distance from $x$ to 0. For example, both 5 and $-5$ are five units away from 0.

It is quite common to use $\varepsilon$, the lower case Greek letter *epsilon*, to denote a "small" positive number. Nowhere will we assume that $\varepsilon$ is small, but the smaller $\varepsilon$ is, the more interesting will be statements involving $\varepsilon$.

Recall the Small Terms Axiom describing what it means to say that the terms $\frac{1}{2}$, $(\frac{1}{2})^2$, $(\frac{1}{2})^3$, $(\frac{1}{2})^4$, ... get small: Given any positive number $\varepsilon$, there is some integer $\ell$ with $(\frac{1}{2})^\ell < \varepsilon$. In our earlier discussion of this axiom, we described a game: you challenge me by giving me a positive number $\varepsilon$; I respond by giving you a term $(\frac{1}{2})^\ell < \varepsilon$. The axiom assumes that every challenge $\varepsilon$ can be met.

Suppose now that a sequence $k_1 < k_2 < k_3 < \ldots < A$ approximates a number $A$ from below. Given $\varepsilon > 0$, we have just seen that there is some number $\ell$ with $(\frac{1}{2})^\ell < \varepsilon$. Since the sequence $k_n$ is increasing, if $n \geq \ell$, then $k_\ell < k_n$, and so $A - k_n < A - k_\ell < (\frac{1}{2})^\ell < \varepsilon$. Therefore, increasing sequences have an extra feature: If $A - k_\ell < \varepsilon$ and $n \geq \ell$, then $A - k_n < \varepsilon$ as well. Thus, in the case of approximation from below, the response we give to a posited $\varepsilon$ meets an even more demanding challenge: not only is $k_\ell < \varepsilon$, but $k_n < \varepsilon$ for all $n \geq \ell$. When we define convergence of arbitrary, not necessarily increasing (or decreasing) sequences, this feature will be made part of the definition.

We now drop the classical assumption that sequences are either increasing or decreasing. After all, not all interesting sequences are monotone; for example, analysis of vibrations necessarily involves sequences and functions that go up and down.

**Definition.** A *tail* of a sequence $\{a_n\} = a_1, a_2, a_3, \ldots$ consists of all the terms of the sequence from some point on; that is, $a_\ell, a_{\ell+1}, a_{\ell+2}, \ldots$ for some number $\ell$.

A tail $a_\ell, a_{\ell+1}, a_{\ell+2}, \ldots$ of a sequence $\{a_n\}$ can be described, therefore, as all those terms $a_n$ with $n \geq \ell$.

The next definition is of fundamental importance; after stating it, we will discuss what it really means.

**Definition.** A sequence $\{a_n\}$ *converges* to a number $L$ if, for every $\varepsilon > 0$, there is some integer $\ell$ with $|a_n - L| < \varepsilon$ for all $n \geq \ell$. In this case, $L$ is called the *limit* of the sequence, and we write $\lim_{n \to \infty} a_n = L$ or, more simply, $a_n \to L$. If a sequence does not converge, one says it *diverges*.

The inequality $|a_n - L| < \varepsilon$ merely says that the distance between $a_n$ and $L$ is "small"; that is, $a_n$ is "close" to $L$.

We can paraphrase this definition. The terms $a_n$ of the sequence are approximations to the limit $L$, and the number $\varepsilon$ can be thought of, negatively, as a mistake; but $\varepsilon$ can also be thought of, more positively, as a tolerance: how far away from the limit are we allowed to stray? Convergence means that no matter what tolerance $\varepsilon$ is specified, all the terms of the sequence, from some point on, are within that tolerance. We shall soon see why we insist that whole tails of a sequence be close to $L$.

Recall that the *open disk* with center $L$ and radius $\varepsilon$ ("open" indicates that we consider only the interior of the disk and not its circular boundary) consists of all the points $X$ in the plane with $d(X, L) < \varepsilon$. By analogy, a "one-dimensional disk" with center $L$ and radius $\varepsilon$ consists of all numbers $x$ with $d(x, L) = |x - L| < \varepsilon$. Thus, this one-dimensional disk is just the open interval $(L - \varepsilon, L + \varepsilon)$ with center $L$ and radius $\varepsilon$. Therefore, the definition of convergence can be restated in more geometric language.

*Geometric Translation*: If every open interval of the form $(L - \varepsilon, L + \varepsilon)$ contains a tail of a sequence $\{a_n\}$, then $a_n \to L$.

The number $\ell$ in the definition depends on $\varepsilon$; if $\varepsilon$ is replaced by a larger number, say, $\varepsilon'$, then the same $\ell$ works; that is, $|a_n - L| < \varepsilon'$ for all $n \geq \ell$. A tail inside of $(L-\varepsilon, L+\varepsilon)$ also lies inside of the larger interval $(L-\varepsilon', L+\varepsilon')$. On the other hand, if $\varepsilon$ is replaced by a smaller positive number $\varepsilon'' < \varepsilon$, then some (larger) integer $\ell$ will probably be needed. If $(L - \varepsilon, L + \varepsilon)$ is replaced by a smaller interval $(L - \varepsilon'', L + \varepsilon'')$, then the tail may have to be replaced by a shorter tail.

*Example*. Let us show that $1/n \to 0$. For any number $B > 1$, there is some integer $\ell > B$ (in courses on the foundations of mathematics, this is called the **Archimedean property**: given any positive numbers $r > s$, there exists an integer $\ell$ with $\ell s > r$ [here, $r = B$ and $s = 1$]). In particular, given $\varepsilon > 0$, there is some integer $\ell > 1/\varepsilon$. Of course, if $n \geq \ell$, then $n \geq \ell > 1/\varepsilon$, so that $1/n < \varepsilon$ for all $n \geq \ell$, by Theorem 1.4(iii). Therefore, $|1/n - 0| = 1/n < \varepsilon$ for all $n \geq \ell$, and so $1/n \to 0$.

*Example*. Here is the reason one defines convergence using tails instead of "small terms." Define a sequence $\{a_n\}$ by $a_{2n-1} = 5$ and $a_{2n} = 1/n$; thus, the sequence begins

$$5, 1, 5, \tfrac{1}{2}, 5, \tfrac{1}{3}, 5, \tfrac{1}{4}, 5, \ldots.$$

On intuitive grounds, the recurrence of 5 every other term should prevent $\{a_n\}$ from converging to 0. The sequence does have small terms, however. We have just seen, for any given $\varepsilon > 0$, that there is $\ell$ with $1/\ell < \varepsilon$; hence, $a_{2\ell} < \varepsilon$, and the sequence has small terms. On the other hand, if $a_n \to 0$, then every open interval $(-\varepsilon, \varepsilon)$ would contain a tail. But if $\varepsilon = 1$, then 5 does not lie in $(-1, 1)$, and so no tail lies within ("half" the terms, those equal to 5, are missing); hence, $a_n \not\to 0$. Exercise 3.33 asks you to prove that this sequence diverges.

Just as in the applications of the Getting Close Principle, one can use the definition of convergence without actually having to calculate, for a given $\varepsilon > 0$, a corresponding integer $\ell$. Almost always, it is enough to know that some integer $\ell$ is available; an explicit calculation of $\ell$ is rarely needed.

**Theorem 3.18.** If $c$ is a number and $a_n \to A$, then $ca_n \to cA$.

*Proof.* If $c = 0$, then $\{ca_n\}$ is the constant sequence of all 0's, and the result is true, by Exercise 3.21.

We may assume, therefore, that $c \neq 0$. Given $\varepsilon > 0$, we must show there is an integer $\ell$ with $|ca_n - cA| < \varepsilon$ whenever $n \geq \ell$. By hypothesis, $a_n \to A$: for any given "error" $\varepsilon'$, there is an integer $\ell'$ with $|a_n - A| < \varepsilon'$ whenever $n \geq \ell'$. In particular, for error $\varepsilon' = \varepsilon/|c|$, there is an integer $\ell'$ with $|a_n - A| < \varepsilon/|c|$ for all $n \geq \ell'$. By Lemma 3.16(iv), $|ca_n - cA| = |c||a_n - A| < |c|(\varepsilon/|c|) = \varepsilon$ for all $n \geq \ell'$, as desired.   •

Some "tinkering" with $\varepsilon$ (e.g., using $\varepsilon/|c|$ instead of $\varepsilon$ to find $\ell$) turns out to be a common routine; we shall tinker again.

**Theorem 3.19.** If $0 < r < 1$ then $r^n \to 0$.

*Remark.* In Exercise 3.30, the hypothesis is weakened to $|r| < 1$.

*Proof.* Let $\varepsilon > 0$ be given. Note that $r < 1$ implies that $1/r > 1$, by Theorem 1.4(iii). Thus, $1/r = 1 + a$, where $a > 0$, and so $(1/r)^n > 1 + na$ for all $n$, by Exercise 1.8. Because $a > 0$, the Archimedean property says there is an integer $\ell$ with $\ell a > 1/\varepsilon$. Now $(1/r)^\ell > 1 + \ell a > \ell a > 1/\varepsilon$, and so $r^\ell < \varepsilon$, by Theorem 1.4(iii). If $n \geq \ell$, then $r^n \leq r^\ell$, and $|r^n - 0| = r^n \leq r^\ell < \varepsilon$ whenever $n \geq \ell$, as desired.   •

We can now prove the Small Terms Axiom.

*Corollary 3.20 (Small Terms Axiom)*. Let $A$ be a positive number. For every $\varepsilon > 0$, there is an integer $\ell$ with $(\frac{1}{2})^n A < \varepsilon$.

*Proof.* Since $0 < \frac{1}{2} < 1$, Theorem 3.19 shows that $(\frac{1}{2})^n \to 0$; setting $c = A$ in Theorem 3.18 now gives $(\frac{1}{2})^n A \to 0$. The definition of convergence says that for every $\varepsilon > 0$, there is some integer $\ell$ with $(\frac{1}{2})^n A = |(\frac{1}{2})^n A - 0| < \varepsilon$ for all $n \geq \ell$. In particular, $(\frac{1}{2})^\ell A < \varepsilon$.  •

The next result shows that the classical Greek notion of approximation is a special case of convergence.

*Theorem 3.21*. If $k_1 < k_2 < k_3 < \ldots < A$ approximates $A$ from below, then $k_n \to A$. Similarly, if $K_1 > K_2 > K_3 > \ldots > A$ approximates $A$ from above, then $K_n \to A$.

*Proof.* We shall prove the result in case the sequence approximates $A$ from below. Let $\varepsilon > 0$. Because $A > k_n$, we have $A - k_n > 0$, and so $|k_n - A| = A - k_n$. By hypothesis, there are inequalities $A - k_n < (\frac{1}{2})^n A$ for every $n \geq 1$. By Corollary 3.20, there is some integer $\ell$ with $(\frac{1}{2})^\ell A < \varepsilon$. Since the sequence $k_n$ is increasing, if $n \geq \ell$, then $k_\ell < k_n$, and so $A - k_n < A - k_\ell < (\frac{1}{2})^\ell < \varepsilon$. Therefore, if $n \geq \ell$,

$$|k_n - A| = A - k_n < \varepsilon,$$

and $k_n \to A$. A similar argument, left to the reader, works for approximation from above.  •

It is now legitimate, for example, to write $\text{area}(P_n) \to \pi r^2$, where $P_n$ is the regular $2^{n+1}$-gon inscribed in a disk of radius $r$ that we constructed when proving the area formula.

We are going to show that convergence is compatible with the standard arithmetic operations, but we first prove a lemma that will help us deal with multiplication and division.

*Lemma 3.22*. If $\{b_n\}$ is a convergent sequence, then there is a number $M$ with $|b_n| < M$ for all $n \geq 1$.

*Proof.* By hypothesis, there is a number $B$ with $b_n \to B$. If $\varepsilon = 1$, there is an integer $\ell$ so that $|b_n - B| < 1$ for all $n \geq \ell$. It follows from the triangle inequality that

$$|b_n| = |b_n - 0| \leq |b_n - B| + |B - 0| < 1 + |B|$$

for all $n \geq \ell$. Now choose $M$ to be the largest of the numbers

$$1 + |b_1|, 1 + |b_2|, \ldots, 1 + |b_{\ell-1}|, 1 + |B|. \quad \bullet$$

**Theorem 3.23.** Assume that $a_n \to A$ and $b_n \to B$.
(i) $a_n \pm b_n \to A \pm B$.
(ii) $a_n b_n \to AB$.
(iii) $(a_n)^m \to A^m$ for all $m \geq 1$.

(Exercise 3.23 shows that convergence is also compatible with division.)

*Proof.* (i) Given $\varepsilon > 0$, there are integers $\ell$ and $\ell'$ so that $|a_n - A| < \varepsilon/2$ for all $n \geq \ell$ and $|b_n - B| < \varepsilon/2$ for all $n \geq \ell'$ (we have tinkered again). For notational convenience, assume that $\ell \geq \ell'$, so that both inequalities hold simultaneously when $n \geq \ell$: the triangle inequality gives

$$\begin{aligned}
|(a_n + b_n) - (A + B)| &= |(a_n - A) + (b_n - B)| \\
&\leq |a_n - A| + |b_n - B| \\
&< \varepsilon/2 + \varepsilon/2 \\
&= \varepsilon.
\end{aligned}$$

Therefore, $a_n + b_n \to A + B$, as desired. The similar argument showing that $a_n - b_n \to A - B$ is left to the reader.

(ii) Given $\varepsilon > 0$, we proceed as we did in our earlier (failed) attempt to prove that $a_1 b_1, a_2 b_2, a_3 b_3, \ldots$ approximates $AB$. Using the triangle inequality and Lemma 3.16(iv),

$$\begin{aligned}
|a_n b_n - AB| &= |a_n b_n - Ab_n + Ab_n - AB| \\
&\leq |a_n b_n - Ab_n| + |Ab_n - AB| \\
&= |a_n - A||b_n| + |A||b_n - B|.
\end{aligned}$$

Again we tinker with $\varepsilon$. There is an integer $\ell$ with $|b_n - B| < \varepsilon/2|A|$ for all $n \geq \ell$. By Lemma 3.22, there is a number $M$ with $|b_n| < M$ for all $n \geq 1$, and so there is an integer $\ell'$ with $|a_n - A| < \varepsilon/2M$ for all $n \geq \ell'$. For notational convenience, we may assume that $\ell \geq \ell'$, so that both inequalities hold simultaneously; whenever $n \geq \ell$,

$$|a_n - A||b_n| + |A||b_n - B| < M \left(\frac{\varepsilon}{2M}\right) + |A| \left(\frac{\varepsilon}{2|A|}\right) = \tfrac{1}{2}\varepsilon + \tfrac{1}{2}\varepsilon = \varepsilon.$$

(iii) We leave this as an exercise for the reader, with the hint that one proceeds by induction on $m \geq 1$; the inductive step will use part (ii) just proved.  •

*Remark.* Parts (i) and (ii) of the theorem can be extended from two sequences to several sequences. It is a routine proof, using induction on $m \geq 2$, that if $a_n^{(1)} \to A^{(1)}, a_n^{(2)} \to A^{(2)}, \ldots$, and $a_n^{(m)} \to A^{(m)}$ [the superscripts $(1)$, $(2), \ldots, (m)$ are just labels describing the first sequence and its limit $A^{(1)}$, the second sequence and its limit $A^{(2)}$, and so forth], then

$$a_n^{(1)} + a_n^{(2)} + \cdots + a_n^{(m)} \to A^{(1)} + A^{(2)} + \cdots + A^{(m)}$$

and

$$a_n^{(1)} a_n^{(2)} \cdots a_n^{(m)} \to A^{(1)} A^{(2)} \cdots A^{(m)}.$$

Of course, part (iii) of the theorem is the special case of the last statement when all the sequences $\{a_n^{(1)}\}, \{a_n^{(2)}\}, \ldots, \{a_n^{(m)}\}$ coincide.

**Theorem 3.24.** Let $f(x) = c_m x^m + \cdots + c_1 x + c_0$ be a polynomial. If $b$ is a number and $a_n \to b$, then $f(a_n) \to f(b)$.

*Proof.* For each fixed $j$ with $1 \leq j \leq m$, we have $(a_n)^j \to b^j$, by Theorem 3.23(iii), and so $c_j(a_n)^j \to c_j b^j$, by Theorem 3.18. The result now follows by the remark after Theorem 3.23.  •

*Definition.* A function $f(x)$ is **continuous** at $b$ if $f(b)$ is defined and, whenever $a_n \to b$ and $f(a_n)$ are all defined, then $f(a_n) \to f(b)$.

*Remark.* There are many (equivalent) definitions of continuity of a function $f(x)$.

Theorem 3.24 says that polynomials are continuous at every $b$; Exercise 3.23 shows that rational functions are continuous whenever the denominator is not zero. Continuity is one of the most important properties a function can have, but we are going to leave further discussion of it to a subsequent course.

The only difference between a *sequence* and a *series* is that one of them has commas and the other one has pluses: a sequence is a list of numbers written as $a_1, a_2, a_3, \ldots$, and a series is a list of numbers written as $a_1 + a_2 + a_3 + \cdots$. Just as one can sometimes associate a number to a sequence (its *limit* if it converges), so one can sometimes associate a number to a series (its *sum* if it converges).

**Definition.** Given a series $a_1 + a_2 + a_3 + \cdots$, define its **partial sums** $s_1, s_2, s_3, \ldots$ inductively as follows:

$$ s_1 = a_1; \text{if } n \geq 1, \text{then } s_{n+1} = s_n + a_{n+1} $$

(of course, $s_n$ is just the sum of the first $n$ terms). If $s_n \to L$, then we say that the series **converges** to its **sum** $L$, and we write

$$ L = \sum_{n=1}^{\infty} a_n. $$

Consider the decimal expansion $\frac{1}{3} = .33333 \cdots$, where the right side is an unending string of 3's. We are looking at a series, for the decimal expansion is really

$$ .3 + .03 + .003 + .0003 + \cdots ; $$

the sequence of partial sums is:

$$ s_1 = 0.3; \ s_2 = 0.33; \ s_3 = 0.333, $$

and, more generally, $s_n = 0.333 \cdots 3$, where there are $n$ 3's. The name *partial sum* is well chosen; one is adding up the first parts of the series.

***Example.*** Consider the geometric series $1 + r + r^2 + r^3 + \cdots$. In Exercise 1.2, we saw that if $r \neq 1$, then

$$
\begin{aligned}
s_n &= 1 + r + r^2 + r^3 + \cdots + r^{n-1} \\
&= \frac{1 - r^n}{1 - r} \\
&= \frac{1}{1 - r} - \frac{r^n}{1 - r}.
\end{aligned}
$$

By Theorem 3.19, $r^n \to 0$ if $0 < r < 1$; it now follows from Theorem 3.18 that $r^n/(1 - r) \to 0$. Finally, Theorem 3.23(i), with $\{a_n\}$ the constant sequence $a_n = 1/(1 - r)$ and $b_n = r^n/(1 - r)$, shows that $s_n \to 1/(1 - r)$. We conclude that

$$
1 + r + r^2 + r^3 + \cdots = \sum_{n=1}^{\infty} r^n = \frac{1}{1 - r}
$$

whenever $0 < r < 1$. (It is shown, in Exercise 3.30, that this result holds whenever $-1 < r < 1$.)

Let us return to

$$
\begin{aligned}
.33333\cdots &= .3 + .03 + .003 + .0003 + \cdots \\
&= \tfrac{3}{10}\left[1 + \tfrac{1}{10} + (\tfrac{1}{10})^2 + (\tfrac{1}{10})^3 + \cdots\right].
\end{aligned}
$$

We recognize a geometric series with $r = \tfrac{1}{10} < 1$. By the example,

$$
1 + \tfrac{1}{10} + (\tfrac{1}{10})^2 + (\tfrac{1}{10})^3 + \cdots = 1/(1 - \tfrac{1}{10}) = 1/(\tfrac{9}{10}) = \tfrac{10}{9}.
$$

Hence, $.33333\cdots = \tfrac{3}{10} \times \tfrac{10}{9} = \tfrac{1}{3}$.

The proof of the next theorem should be compared with the proof of Theorem 3.3, which is the same result in the classical vein.

**Theorem 3.25.** If $B = .99999\ldots$, then $B = 1$.

*Proof.* By definition, $B$ is the sum of the series

$$B = 9r + 9r^2 + 9r^3 + \cdots = 9r(1 + r + r^2 + \cdots),$$

where $r = \frac{1}{10}$. As we have just shown above, the sum of the geometric series $1 + r + r^2 + r^3 + \cdots$ is $\frac{10}{9}$, so that $B = 9r\frac{10}{9} = 9 \cdot \frac{1}{10} \cdot \frac{10}{9} = 1$.  ●

We have been terribly unfair to the Greeks. After all, they set out to prove the area and circumference formulas for disks, and they succeeded brilliantly by inventing a notion of limit appropriate to the problems. How, then, can we fault them for not anticipating that their descendents, two thousand years in the future, would invent calculus and need a more sophisticated notion of limit? I confess that my goal in this chapter is to use the Greek notion of approximation to promote the modern notion of limit. Even though the ancient definition is subtle, it is simpler than the contemporary one full of quantifiers; moreover, the ancient notion arises in a geometric context: one can see polygons approximating a disk. This chapter was designed, however, not to praise the ingenuity of our ancestors, but to let their ideas serve to help us understand the modern perfecting of them.

### Exercises

**3.21.** Let $a_1, a_2, a_3, \ldots$ be a constant sequence; that is, there is some number $c$ with $a_n = c$ for all $n$. Prove that $a_n \to c$.

**3.22.** Prove that $|x - L| < \varepsilon$ if and only if $L - \varepsilon < x < L + \varepsilon$; that is, $x$ lies in the open interval $(L - \varepsilon, L + \varepsilon)$. [Hint: Think of $(L - \varepsilon, L + \varepsilon)$ as a "one-dimensional disk."]

**3.23.** (i) If $b_n \to B \neq 0$ and all $b_n \neq 0$, prove that there is an integer $\ell$ and some number $N$ with $1/|b_n| < N$ for all $n \geq \ell$. [Hint: In the definition of $b_n \to B$, let $\varepsilon = |B|/2$, let $\ell$ determine the corresponding tail, and define $N = 2/|B|$.]
(ii) Assume that $a_n \to A$ and $b_n \to B$. Prove that if $B \neq 0$ and all $b_n \neq 0$, then $a_n/b_n \to A/B$.
(iii) Let $f(x)$ and $g(x)$ be polynomials. If $g(b) \neq 0$, prove that the rational function $f(x)/g(x)$ is continuous at $b$.

**3.24.** Prove that the sequence $\{a_n\}$ given by $a_n = (-1)^n$ diverges.

**3.25.** Prove that a convergent sequence has only one limit: if $a_n \to A$ and $a_n \to L$, then $A = L$. (Compare Theorem 3.4.)

**3.26.** (i) Let $a_1, a_2, a_3, \ldots$ be a sequence with $a_n \to L$, and consider the new sequence $a_2, a_3, a_4, \ldots$ obtained by eliminating the first term. Show that the new sequence also converges to $L$.

(ii) Suppose one deletes (possibly infinitely many) terms of a sequence $\{a_n\}$. If infinitely many terms remain, they are called a ***subsequence*** of $\{a_n\}$ (for example, we may delete finitely many terms, or we may delete every third term). Show that if $a_n \to L$, then every subsequence of $\{a_n\}$ also converges to $L$. (Hint: If $b_1, b_2, b_3, \ldots$ is a subsequence of $\{a_n\}$, then each $b_k = a_{n(k)}$ with $n(k) \geq k$.)

**3.27.** (***Sandwich theorem***). Suppose that $\{a_n\}$, $\{b_n\}$, and $\{c_n\}$ are sequences with $a_n \to L, b_n \to L$, and $a_n \leq c_n \leq b_n$ for all $n \geq 1$. Prove that $c_n \to L$. (Compare Theorem 3.5.)

**3.28.** (i) Show that $1 - (\frac{2}{3})^n \to 1$.

(ii) Show that $(n - 1)/n \to 1$.

**3.29.** Prove that $a_n \to A$ if and only if $A - a_n \to 0$.

**3.30.** (i) Prove that if $|r| < 1$, then $r^n \to 0$.

(ii) Prove that the geometric series $\sum_{n=0}^{\infty} r^n$ converges, with sum $1/(1 - r)$, whenever $|r| < 1$.

**3.31.** Prove, using the definition of convergence, that $\frac{1}{3} = .3333\cdots$.

**3.32.** Assume that there are two sequences converging to the same limit: $a_n \to A$ and $b_n \to A$. Prove that

$$a_1, b_1, a_2, b_2, a_3, b_3, a_4, b_4, \ldots \to A.$$

**3.33.** Show that the sequence $\{a_n\}$ defined by $a_{2n-1} = 5$ and $a_{2n} = 1/n$ diverges. (Comparison with Exercise 3.32 shows that one must assume, in that exercise, that both sequences converge to the same limit.)

**3.34.** Define a sequence $\{a_n\}$ by $a_{2n-1} = \frac{1}{n}$ and $a_{2n} = (\frac{1}{2})^n$. The sequence begins:
$1, \frac{1}{2}, \frac{1}{2}, \frac{1}{4}, \frac{1}{3}, \frac{1}{8}, \frac{1}{4}, \frac{1}{16}, \cdots$.

(i) Prove that $a_n \to 0$.

(ii) Prove that the terms $a_n$ do not get "closer and closer" to the limit 0; that is, if $n > \ell$, then it does not follow that $|a_n - 0| < |a_\ell - 0|$.

**3.35.** (i) Earlier, we constructed a sequence $\{P_n\}$ of $2^{n+1}$-gons inscribed in a disk of radius $r$, as well as a sequence $\{Q_n\}$ of circumscribed $2^{n+1}$-gons. If $p_n$ is the perimeter of $P_n$, and if $q_n$ is the perimeter of $Q_n$, prove that $p_n \to 2\pi r$ and $q_n \to 2\pi r$. (See the proof of Theorem 3.13.)

(ii) Prove that $\sin(\pi/2^n)/(\pi/2^n) \to 1$. (This is a version of $\lim_{x\to 0} \sin x/x = 1$.) You may not use L'Hôpital's Rule.

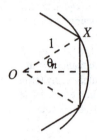

Figure 3.23

**3.36.** (i) Prove that if $a_n$ lies in the open interval $(b - \frac{1}{n}, b + \frac{1}{n})$ for all $n \geq 1$, then $a_n \to b$. [Hint: Either use the definition of convergence or the Sandwich theorem (Exercise 3.22).]

(ii) Let $f(x)$ be a polynomial, and assume that $b$ is not a root of $f(x)$, i.e., $f(b) \neq 0$. Prove that there is some interval $(b - \varepsilon, b + \varepsilon)$ so that $f(a) \neq 0$ for all $a$ in $(b - \varepsilon, b + \varepsilon)$. [Hint: Use part (i).]

**3.37.** Assume that $\{a_n\}$ is a decreasing sequence; that is, $a_n \geq a_{n+1}$ for all $n \geq 1$. If $a_n \to t$, prove that $a_n \geq t$ for all $n \geq 1$.

# Chapter 4

# Polynomials

Stand firm in your refusal to remain conscious during algebra. In real life, I assure you, there is no such thing as algebra.

*Fran Lebowitz*

A *quadratic polynomial* is an expression of the form

$$f(x) = ax^2 + bx + c,$$

where $a$, $b$, and $c$ are given numbers, with $a \neq 0$, called its *coefficients*. A *root* (or *zero*) of $f(x)$ is a number $r$ such that

$$f(r) = ar^2 + br + c = 0.$$

The Babylonians knew, 4000 years ago, how to find the roots of quadratics. One example of a Babylonian problem is, given a (positive) number $b$, to find a number $x$ such that

$$x + \frac{1}{x} = b;$$

it is easy to see that $x$ is a root of the quadratic

$$x^2 - bx + 1.$$

Quadratic polynomials arise in geometric problems. For example, finding the intersection of a line and a circle leads to a quadratic equation, as we saw in our discussion of the method of Diophantus. Another example occurs in finding the sides $x$ and $y$ of a rectangle given its area $A$ and perimeter $p$. The equations

$$xy = A \quad \text{and} \quad 2(x + y) = p$$

yield

$$2\left(x + \frac{A}{x}\right) = p,$$

and this simplifies to give the quadratic equation

$$2x^2 - px + 2A = 0.$$

Of course, once $x$ has been found, then $y = A/x$ is also determined.

Quadratic polynomials arise in applied problems. For example, around 1604, Galileo (1564–1643) performed his famous experiment at the leaning tower in Pisa. He found that two objects of different weights fall equally fast, so that their common velocity $v = v(t)$ depends only on the duration $t$ of their fall and not upon their masses; indeed, $v(t) = gt$, where the *gravitational constant* $g$ is approximately 32 if distance is measured in feet and time is measured in seconds. Galileo was able to use this fact to describe the path

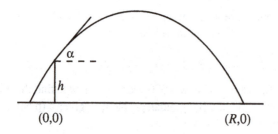

Figure 4.1

of a projectile. If a cannon on a hill $h$ feet above the ground makes an angle of $\alpha$ degrees with the ground, then Galileo showed that a shell fired with initial velocity $v_0$ is, after $t$ seconds, at the point $(x, y)$, where

$$x = v_0 t \cos \alpha \qquad (1)$$
$$y = -\tfrac{1}{2}gt^2 + v_0 t \sin \alpha + h \qquad (2)$$

For example, suppose that a cannon on a hill 32 feet above the ground makes an angle $\alpha = 30°$ with the ground, and that the initial velocity $v_0$ of a fired shell is 160 feet per second. The position of the shell $t$ seconds after firing is

$$(x, y) = (160(\tfrac{\sqrt{3}}{2})t, -16t^2 + 160\tfrac{1}{2}t + 32)$$
$$= (80\sqrt{3}t, -16t^2 + 80t + 32).$$

What is the horizontal distance $R$ traveled by the shell when it hits the ground? Since the $y$-coordinate of $(x, y)$ is the height above the ground, $R$ is the $x$-coordinate of the point for which $y = 0$. This gives a quadratic equation (after dividing by $-16$)

$$t^2 - 5t - 2 = 0;$$

the corresponding value of $t$ is the number of seconds it takes the shell to hit the ground. Once we know this time, we can find the distance $R$ using Eq. (1). The quadratic formula gives

$$t = \tfrac{1}{2}(5 + \sqrt{33}) \quad \text{and} \quad t = \tfrac{1}{2}(5 - \sqrt{33}).$$

The second root $\tfrac{1}{2}(5 - \sqrt{33})$, being negative, is ignored, for negative time occurs before the firing. Thus, $t = \tfrac{1}{2}(5 + \sqrt{33}) \approx 5.37$ and $R = v_0 t \cos \alpha = 160(\tfrac{\sqrt{3}}{2}) \times 5.37 \approx 744$ feet. The algebraic discussion led to two solutions, but only one of them was relevant, and so we ignored the other.

Let us begin our discussion of quadratics by proving an elementary fact of arithmetic often introduced in high schools as a mysterious revelation from some ineffable omniscient source: $(-1) \times (-1) = +1$. One of the standard arithmetic properties is the familiar **distributive law**: for every three numbers $a$, $b$, and $c$, perhaps negative, we have

$$(a + b)c = ac + bc.$$

Reading from left to right, distributivity says that $c$ can be "multiplied through" $a + b$ to get $ac + bc$; reading from right to left, it says that $c$ can be "factored out" of $ac + bc$ to get $(a + b)c$.

Let us first see, for any number $a$, that $0 = 0 \times a$. After all,

$$0 \times a = (0 + 0) \times a = (0 \times a) + (0 \times a).$$

Subtracting $0 \times a$ from both sides gives $0 = 0 \times a$.

Let us apply the distributive law with $a = -1, b = 1$, and $c = -1$; we have

$$
\begin{aligned}
0 = 0 \times (-1) &= (-1 + 1)(-1) \\
&= [(-1) \times (-1)] + [1 \times (-1)] \qquad \text{(distributive law)} \\
&= [(-1) \times (-1)] + (-1).
\end{aligned}
$$

Adding 1 to both sides gives $1 = (-1) \times (-1)$, as desired.

*Remark* 1. Dividing a number $m$ by a number $n$ yields the number $m/n$, and $n \times (m/n) = m$; that is, dividing and multiplying are opposite in the sense that one operation undoes the other. For example, dividing 24 by 3 gives 8, and multiplying 8 by 3 gives back 24: $3 \times (24/3) = 24$. This is the basic reason why one can never divide by 0. If, for example, $24/0$ made sense, then $0 \times (24/0) = 24$. But we have just shown that $0 \times a = 0$ for every number $a$; in particular, if $24/0$ were a number, then $0 \times (24/0) = 0 \neq 24$. We are forced to conclude that $24/0$ is not a number; that is, one cannot divide by 0.

*Remark* 2. This is an appropriate place to say a word about the symbol $\infty$ (infinity). There are times when $\infty$ behaves as a number, but there are times when it does not. For example, we have just proved that $0 \times a = 0$ for every number $a$, whereas L'Hôpital's Rule tells us that $0 \times \infty$ may or may not be 0. Perhaps the worst type of restaurant is an inconsistent one: sometimes the food is wonderful; sometimes it is terrible; but one never knows, in advance, which type of meal one will get. So, too, with $\infty$. If one cannot predict at the outset whether the outcome is correct, then one cannot safely use the result. It is best, therefore, to use $\infty$ only as a convenient notation.

We now show that $p^2 \geq 0$ for every real number $p$. Of course, this is true if $p \geq 0$; if $p$ is negative, then $p = -q$ for some positive number $q$. Thus, $p = (-1)q$, and $p^2 = [(-1)q]^2 = (-1)^2 q^2 = q^2 \geq 0$. It follows that if $p \neq 0$, then there is no (real) number $r$ with $r^2 = -p^2$; that is, the quadratic $x^2 + p^2$ has no (real) roots.

**Theorem 4.1 (Quadratic Formula).** If $a \neq 0$, the roots of a quadratic $f(x) = ax^2 + bx + c$ are given by

$$x = \frac{-b \pm \sqrt{b^2 - 4ac}}{2a}.$$

provided its **discriminant** $b^2 - 4ac \geq 0$. If $b^2 - 4ac$ is negative, then the equation has no (real) roots.

*Proof.* The idea is to "complete the square." The roots of

$$ax^2 + bx = -c,$$

are the same as the roots of

$$x^2 + \frac{b}{a}x = -\frac{c}{a}.$$

To make the left side look like a square $(x + d)^2 = x^2 + 2dx + d^2$, add $b^2/4a^2$ to both sides of the equation to obtain

$$x^2 + \left(\frac{b}{a}\right)x + \frac{b^2}{4a^2} = -\frac{c}{a} + \frac{b^2}{4a^2}.$$

The equation can now be rewritten as:

$$\left(x + \frac{b}{2a}\right)^2 = \frac{b^2 - 4ac}{4a^2}.$$

For any real number $x$, the left side, being a square, is never negative. If the discriminant $b^2 - 4ac < 0$, then the right side is negative, and so there are no (real) roots. If $b^2 - 4ac \geq 0$, however, then one can take the square root of both sides to get

$$x + \frac{b}{2a} = \frac{\pm\sqrt{b^2 - 4ac}}{2a},$$

and so

$$x = -\frac{b}{2a} \pm \frac{\sqrt{b^2 - 4ac}}{2a} = \frac{-b \pm \sqrt{b^2 - 4ac}}{2a}. \quad \bullet$$

The phrase "completing the square" can be taken literally. Given a quadratic $x^2 + ux = v$ with $u \geq 0$, view $x^2 + ux$ as the area pictured in

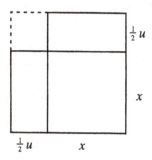

Figure 4.2

Figure 4.2 (compare Figure 2.16). One completes the square by adding on the corner square having area $\frac{1}{4}u^2$. The area of the large square is $(x + \frac{1}{2}u)^2$; if $v + \frac{1}{4}u^2 \geq 0$, then there is a square with side length $x + \frac{1}{2}u$ and area $v + \frac{1}{4}u^2$. In terms of the quadratic $x^2 + ux - v$, the geometric square can be completed if (its discriminant) $u^2 + 4v \geq 0$.

**Corollary 4.2.** A quadratic $f(x) = ax^2 + bx + c$ has a unique root if and only if its discriminant $b^2 - 4ac = 0$.

*Proof.* If $b^2 - 4ac = 0$, then the quadratic formula gives $x = (-b \pm 0)/2a = -b/2a$ as the only root of $f(x)$.

Conversely, if $b^2 - 4ac \neq 0$, then the quadratic formula displays two distinct roots if $b^2 - 4ac > 0$ and no roots if $b^2 - 4ac < 0$.  •

Here is a variant of the quadratic formula that is useful in some computations. Consider the quadratic

$$f(x) = 10^{-20}x^2 - 3x - 7.$$

The quadratic formula gives the roots

$$r_1 = \frac{3 + \sqrt{9 + 28 \times 10^{-20}}}{2 \times 10^{-20}} \quad \text{and} \quad r_2 = \frac{3 - \sqrt{9 + 28 \times 10^{-20}}}{2 \times 10^{-20}}.$$

Because $28 \times 10^{-20}$ is very small, $9 + 28 \times 10^{-20}$ is very close to 9, and so the radical $\sqrt{9 + 28 \times 10^{-20}}$ is very close to but slightly larger than $3 = \sqrt{9}$. Thus, $r_1$ is positive and $r_2$ is negative. Now

$$3 + \sqrt{9 + 28 \times 10^{-20}}$$

is approximately 6, and so $r_1 \approx 6/(2 \times 10^{-20}) = 3 \times 10^{20}$. On the other hand, one needs a very good calculator to estimate the negative root $r_2$ because the denominator $3 - \sqrt{9 + 28 \times 10^{-20}}$ is so small.

There is an alternate version of the quadratic formula that allows estimation of $r_2$. By Theorem 4.1, the roots of a quadratic $ax^2 + bx + c$ are

$$r_1 = \frac{-b + \sqrt{b^2 - 4ac}}{2a} \quad \text{and} \quad r_2 = \frac{-b - \sqrt{b^2 - 4ac}}{2a}.$$

Denote the discriminant $b^2 - 4ac$ by $d$, and rationalize the numerator (!):

$$r_2 = \frac{-b - \sqrt{d}}{2a}$$
$$= \frac{-b - \sqrt{d}}{2a} \cdot \frac{-b + \sqrt{d}}{-b + \sqrt{d}}$$
$$= \frac{b^2 - d}{2a(-b + \sqrt{d})}.$$

But $d = b^2 - 4ac$, so that

$$r_2 = \frac{4ac}{2a(-b + \sqrt{b^2 - 4ac})}$$
$$= \frac{-2c}{b - \sqrt{b^2 - 4ac}}.$$

Similarly,

$$r_1 = \frac{-2c}{b + \sqrt{b^2 - 4ac}}.$$

This version of the quadratic formula allows us to estimate the negative root in the example above. The rewritten formula gives

$$r_2 = \frac{14}{-3 - \sqrt{9 + 28 \times 10^{-20}}};$$

the denominator is approximately $-6$, and $r_2 \approx 14/(-6) \approx -2.333$.

**Theorem 4.3**. If $f(x) = ax^2 + bx + c$ is a quadratic with $a > 0$, then $f(x) \geq 0$ for all real numbers $x$ if and only if $b^2 \leq 4ac$.

*Proof.* If $f(x) \geq 0$ for all $x$, then

$$
\begin{aligned}
0 \leq f(-b/2a) &= a(-b/2a)^2 + b(-b/2a) + c \\
&= b^2/4a - b^2/2a + c \\
&= (-b^2 + 4ac)/4a.
\end{aligned}
$$

Since $a > 0$, we must have $0 \leq -b^2 + 4ac$; that is, $b^2 \leq 4ac$.

Conversely, if $b^2 \leq 4ac$, then the discriminant $b^2 - 4ac \leq 0$, and so the graph of $y = f(x)$ cannot cross the $x$-axis; that is, its graph lies either entirely above or entirely below it (if $b^2 - 4ac = 0$, the graph touches the $x$-axis in one point). In the first case, $f(x) \geq 0$ for all real numbers $x$; in the second case, $f(x) \leq 0$ for all real numbers $x$. Let us decide which possibility occurs here. Now $f(0) = c$. Because $0 \leq b^2 < 4ac$, both $a$ and $c$ must have the same sign. By hypothesis, $a > 0$, so that $f(0) = c > 0$. Therefore, $f(x) \geq 0$ for all $x$.   •

The following result was found by A. L. Cauchy (1789–1857).

**Theorem 4.4 (Cauchy's Inequality).** For any numbers $a_1, a_2, \ldots, a_n, b_1, b_2,$ $\ldots, b_n$, there is an inequality

$$
(a_1 b_1 + a_2 b_2 + \cdots + a_n b_n)^2 \leq (a_1^2 + a_2^2 + \cdots + a_n^2)(b_1^2 + b_2^2 + \cdots + b_n^2).
$$

Moreover, equality holds if and only if there is a nonzero number $u$ with $b_i = a_i u$ for all $i = 1, 2, \ldots, n$.

*Proof.* If we define

$$
f(x) = (a_1 x + b_1)^2 + (a_2 x + b_2)^2 + \cdots + (a_n x + b_n)^2,
$$

then expanding and collecting terms gives

$$
f(x) = Ax^2 + Bx + C,
$$

where $A = a_1^2 + a_2^2 + \cdots + a_n^2$, $B = 2(a_1b_1 + a_2b_2 + \cdots + a_nb_n)$, and $C = b_1^2 + b_2^2 + \cdots + b_n^2$. The original version of $f(x)$ as a sum of squares shows that $f(x) \geq 0$ for all real numbers $x$, and so Theorem 4.3 (which applies because $A > 0$) gives $B^2 \leq 4AC$; since $B^2 = 4(a_1b_1 + a_2b_2 + \cdots + a_nb_n)^2$, canceling 4 from each side gives Cauchy's inequality.

Assume that $b_i = a_i u$ for all $i$. If

$$f(x) = (a_1 x + b_1)^2 + \cdots + (a_n x + b_n)^2 = Ax^2 + Bx + C$$

(using the notation above), then $f(r) = 0$, for some number $r$, if and only if $a_i r + b_i = 0$ for all $i$. Hence, $r = -u$ is the only root of $f(x)$, and Corollary 4.2 says that the discriminant of $f(x)$ is 0; that is, $B^2 = 4AC$.

Conversely, if $B^2 = 4AC$, then Corollary 4.2 says that $f(x)$ has only one root. But a number $r$ is a root of $f(x) = (a_1 x + b_1)^2 + (a_2 x + b_2)^2 + \cdots + (a_n x + b_n)^2$ if and only if $a_i r + b_i = 0$ for all $i$; that is, $b_i = a_i u$ for all $i$, where $u = -r$.  •

There is a geometric interpretation of Cauchy's inequality when $n = 3$ (after one takes the square root of both sides): If $v = (a_1, a_2, a_3)$ and $w = (b_1, b_2, b_3)$ are vectors, then their *dot product* $v \cdot w = a_1b_1 + a_2b_2 + a_3b_3$, and Cauchy's inequality says that $|v \cdot w| \leq \|v\| \|w\|$, where $\|v\|$ denotes the length of $v$ (indeed, for any $n \geq 1$, one can define vectors $v = (a_1, a_2, \ldots, a_n)$ and $w = (b_1, b_2, \ldots, b_n)$ and a dot product $v \cdot w = a_1b_1 + a_2b_2 + \cdots + a_nb_n$, and this geometric interpretation remains valid). Moreover, the theorem adds that there is equality, $|v \cdot w| = \|v\| \|w\|$, if and only if $w$ is a scalar multiple of $v$.

The following problem arises in Statistics. If one regard points $(s_1, t_1)$, $(s_2, t_2), \ldots, (s_n, t_n)$ in the plane as data, can one find the simplest curve that "fits" them? The reason for the quotation marks around the word *fits* is that there are different ways to describe what one means by this intuitive notion. The next theorem answers this problem when the curve to be found is a straight line through the origin. One reasonable measure of the goodness of fit of a line $y = mx$ to the points is the number

$$D(m) = \sum_{i=1}^{n} (t_i - ms_i)^2.$$

Notice that $D(m) \geq 0$, and that our discussion above shows that $D(m) = 0$ if and only if $t_i - ms_i = 0$ for all $i$; that is, if all the points $(s_i, t_i)$ lie on the line $y = mx$.

*Corollary 4.5 (Least Squares).* Given $n$ points $(s_1, t_1), (s_2, t_2), \ldots, (s_n, t_n)$, let

$$A = s_1^2 + s_2^2 + \cdots + s_n^2 \quad \text{and} \quad B = 2(s_1 t_1 + s_2 t_2 + \cdots + s_n t_n).$$

Then the line $\ell$ with equation $y = (B/2A)x$ best fits the $n$ points in the sense that of all the lines $y = mx$ through the origin, $\ell$ minimizes

$$D(m) = \sum_{i=1}^{n} (t_i - ms_i)^2.$$

*Proof.* As in the proof of Theorem 4.4, we may write

$$D(m) = Am^2 - Bm + C,$$

where $A$ and $B$ are defined as in the statement (so that $A$ is positive), and $C = t_1^2 + t_2^2 + \cdots + t_n^2$. As in the proof of Theorem 4.1, completing the square gives the equation

$$\frac{D(m) - C}{A} + \frac{B^2}{4A^2} = \left( m - \frac{B}{2A} \right)^2,$$

so that

$$D(m) = A \left( m - \frac{B}{2A} \right)^2 + \frac{4AC - B^2}{4A^2}.$$

Because $m$ is not involved in the term $(4AC - B^2)/4A^2$, the number $D(m)$ is smallest when $A(m - B/2A)^2$ is smallest; but $A(m - B/2A)^2 \geq 0$, and it is 0 when $m = B/2A$.  •

Using linear algebra, one can fit data with a line that need not go through the origin. Indeed, for any $d \geq 1$, one can find a polynomial of degree $d$ whose graph "best fits" the data.

## *Exercises*

**4.1.** If $r_1$ and $r_2$ are the roots of $f(x) = x^2 + bx + c$, compute $b$ and $c$ in terms of $r_1$ and $r_2$. [Hint: $f(x) = (x - r_1)(x - r_2)$.]

**4.2.** If $a, b,$ and $c$ are odd integers, prove that $ax^2 + bx + c$ has no rational roots.

**4.3.** Find the points where the line with equation $y = 2x + 2$ intersects the circle of radius 1 and center $(0, 0)$.

**4.4.** Suppose that a rectangle having sides of lengths $x$ and $y$ has area $A$ and perimeter $p$. If $p^2 - 16A \geq 0$, show that the roots of $2u^2 - pu + 2A$ are $x$ and $y$.

**4.5.** Find and estimate the roots of $10^{-50}x^2 + 2x - 9$.

**4.6.** Suppose a cannon on the ground makes an angle of $45°$.
(i) If the initial velocity of a shell is $v_0 = 80$ feet per second, what is the horizontal distance $R$ the shell travels when it hits the ground?
(ii) What should the initial velocity $v_0$ of a shell be in order that it hit a target 400 feet away?

*Remark.* In most high school algebra classes, one is hounded to "rationalize the denominator"; for example, one must rewrite $\frac{1}{\sqrt{2}}$ as $\frac{\sqrt{2}}{2}$. The commandment — no radicals below the belt — originated before the time of easy accessibility to electronic calculators. It is a good idea to have some idea of the magnitude of a number; one knows that $\sqrt{2} \approx 1.414 \cdots$, and so $\frac{\sqrt{2}}{2} \approx .707$. On the other hand, it is not obvious at a glance what the approximate value of $\frac{1}{1.414\cdots}$ is. Though it is often convenient to write $\frac{\sqrt{2}}{2}$, it is sometimes better to write $\frac{1}{\sqrt{2}}$, as in the calculations for Exercise 4.6. One should not be rigid.

**4.7.** Prove, for any numbers $a_1, a_2, \ldots, a_n$, that

$$\left( \frac{a_1 + a_2 + \cdots + a_n}{n} \right)^2 \leq \frac{a_1^2 + a_2^2 + \cdots + a_n^2}{n}.$$

## COMPLEX NUMBERS

"Reeling and Writhing, of course, to begin with," the Mock Turtle replied, "and the different branches of Arithmetic—Ambition, Distraction, Uglification, and Derision."

*Lewis Carroll*

Solving quadratic equations involves addition, subtraction, multiplication, and division, as well as extracting square roots. Although the first four operations are "rational", that is, starting with rational numbers, they produce only rational numbers, this is not the case for the square root operation. This is why roots of quadratics with rational coefficients, e.g., $x^2 - 2$, may not be rational numbers, or why there may be no roots at all. If we could take square roots of negative real numbers (using some vaster concept of number), then the quadratic formula would be valid without restriction.

We must say at once that the discovery of complex numbers was not at all motivated by a compulsion to provide roots where none had existed before. In the Middle Ages and before, most quadratic equations arose from geometric problems, as we have seen, and the desired root, being a length of a line segment, for example, was known to be a positive real number. Thus, even real roots were ignored if they happened to be negative, for no line segment is $-2$ feet long! We did this very thing ourselves when we ignored the negative root when discussing the distance traveled by a shell fired from a hill. If one could ignore negative roots, one could surely ignore complex roots.

Complex numbers were first studied seriously in the sixteenth century, when mathematicians sought to understand roots of cubics. Later in this chapter, we will discuss a formula giving the roots of

$$x^3 + bx^2 + cx + d.$$

One aspect of this formula, as we shall see, is that it may give a real positive root in terms of square roots of negative numbers. Since such a positive real root cannot be ignored, how can one ignore complex numbers?

The complex numbers form a larger system of numbers containing the usual numbers (which, in contrast, are called "real numbers"). The terms "real" and "imaginary" are relics of the first tentative steps taken by sixteenth century mathematicians trying to understand a new idea that had been thrust upon them. Just as the problem of doubling the square led to the study of irrational numbers, so the cubic formula led to the study of complex numbers.

***Definition.*** A ***complex number*** is an ordered pair $z = (a, b)$ of real numbers.

If $z = (a, b)$, one often calls $a$ the ***real part*** of $z$ and $b$ the ***imaginary part*** of $z$. Beware: the imaginary part of $z$ is the real number $b$.

We have just defined complex numbers as points in the plane. One consequence of this definition is that two complex numbers $(a, b)$ and $(c, d)$ are ***equal*** precisely when their real parts are equal: $a = c$, and their imaginary parts are equal: $b = d$. One equation between complex numbers is the same as two equations between real numbers.

The point $(x, 0)$ on the $x$-axis is identified with the real number $x$. Thus, real numbers are just special complex numbers; in the phrase "let $z$ be a complex number," $z$ might be a real number.

We now extend the four arithmetic operations from the real numbers to the complex numbers.

***Ambition.*** The ***sum*** $(a, b) + (c, d)$ of two complex numbers is the complex number $(a + c, b + d)$.

Since $a, b, c,$ and $d$ are real numbers, $a + c$ and $b + d$ are just ordinary sums. Notice that if $(a, b)$ and $(c, d)$ are real numbers, that is, if $b = 0 = d$, then we get nothing new: $(a, 0) + (c, 0) = (a + c, 0)$. Thus, addition of complex numbers agrees with addition of real numbers when it can.

There is a geometric interpretation of complex addition: the points $(0, 0), (a, b), (c, d),$ and $(a + c, b + d)$ form a parallelogram.

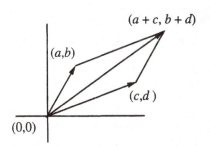

Figure 4.3

We extend multiplication from real numbers to complex numbers in two stages: first: real × complex, and then: complex × complex. Given a complex number $(a, b)$ and a real number $r = (r, 0)$, define

$$r(a, b) = (ra, rb).$$

The usual rules of arithmetic hold in this new context. For example, for every complex number $(a, b)$, we have $0(a, b) = (0, 0)$ and $1(a, b) = (a, b)$.

There is a geometric interpretation of this multiplication. Regard a point $P = (a, b)$ as an arrow from $(0, 0)$ to $P$.

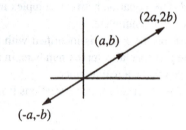

Figure 4.4

Multiplying by 2 doubles the length of the arrow, and more generally, multiplying $(a, b)$ by a positive number $r$ stretches the arrow by a factor of $r$. Multiplying by $r = -1$ changes $(a, b)$ to $(-a, -b)$, thus reversing the direction of the original arrow. Finally, multiplying $(a, b)$ by $-r$ (where $r$ is positive) reverses the arrow and stretches it by a factor of $r$. Because stretching an arrow is a change of scale, this multiplication of complex numbers by real numbers is often called *scalar multiplication*.

*Distraction*. The **difference** of complex numbers is defined by

$$(a, b) - (c, d) = (a, b) + (-1)(c, d)$$

$$= (a, b) + (-c, -d)$$

$$= (a - c, b - d).$$

*Definition*. The complex number $(0, 1)$ is denoted by $i$.

The $x$-axis is often called the real axis, and the $y$-axis is often called the imaginary axis. I once heard the following message on a telephone answering machine. "The number you have reached is imaginary; please rotate your telephone 90°."

***Theorem 4.6.*** Every complex number $z = (a, b)$ is equal to $a + bi$, where $a$ and $b$ are real.

*Proof.* We have

$$(a, b) \;=\; (a, 0) + (0, b)$$

$$=\; a(1, 0) + b(0, 1)$$

$$=\; a + bi,$$

for $(1, 0)$ is the real number 1. •

From now on, we shall use this standard notation for complex numbers.

Addition of complex numbers behaves very much like addition of real numbers.

***Theorem 4.7.*** Let $u$, $v$, and $w$ be complex numbers.
(i)  $u + v = v + u$.
(ii)  $(u + v) + w = u + (v + w)$.
(iii)  $u + 0 = u$.
(iv)  $u + (-u) = 0$.

*Proof.* If $u = a + bi$ and $v = c + di$, then

$$u + v = (a + c) + (b + d)i \quad \text{and} \quad v + u = (c + a) + (d + b)i.$$

Since $a + c = c + a$ and $b + d = d + b$ (for this is a standard property of ordinary addition of real numbers), the real parts agree and the imaginary parts agree; hence, $u + v = v + u$. Verification of the other parts is also routine. •

The product of any two complex numbers has not yet been defined; we have only given the product of a real number and a complex number.

***Uglification.*** Define $i^2 = -1$ The ***product*** of any two complex numbers is now forced upon us:

$$(a + bi)(c + di) \;=\; ac + adi + bci + bdi^2$$

$$=\; (ac - bd) + (ad + bc)i.$$

There is no need to memorize this formula; just multiply as you ordinarily would, and then recall that $i^2 = -1$.

Complex multiplication has many of the properties of real multiplication (notice that multiplication of complex numbers agrees with the usual multiplication whenever both factors are real numbers).

**Theorem 4.8.** Let $u$, $v$, and $w$ be complex numbers.
(i) $uv = vu$.
(ii) $u(vw) = (uv)w$.
(iii) $u(v + w) = uv + uw$.

*Proof.* All are routine, using the corresponding property of multiplication of real numbers. We prove only (iii), the distributive law, leaving the proofs of (i) and (ii) to the reader. Let $u = a + bi$, $v = c + di$, and $w = e + fi$. Now

$$u(v + w) = (a + bi)[(c + e) + (d + f)i]$$

$$= [a(c + e) - b(d + f)] + [a(d + f) + b(c + e)]i.$$

On the other hand,

$$uv + uw = (a + bi)(c + di) + (a + bi)(e + fi)$$
$$= [ac - bd + (ad + bc)i] + [ae - bf + (af + be)i].$$

The real part of each is $ac + ae - bd - bf$, and the imaginary part of each is $ad + af + bc + be$. Hence, $u(v + w) = uv + uw$.   •

It follows from part (i) that $bi = ib$, and so

$$a + bi = a + ib.$$

**Definition.** The *conjugate* of a complex number $z = a + bi$, denoted by $\bar{z}$, is defined by

$$\bar{z} = a - bi.$$

If $z = a + bi$, then $z\bar{z} = (a + bi)(a - bi) = a^2 - b^2i^2 = a^2 + b^2$. Thus, $z\bar{z}$ is always a nonnegative real number; indeed, if $z \neq 0$, then $z\bar{z} > 0$.

**Theorem 4.9.** If $u \neq 0$, then there is a complex number $w$ with $uw = 1$ (of course, one writes $w = 1/u$).

*Proof.* Let $u = a + bi$. Because $u \neq 0$, we have $u\bar{u} = a^2 + b^2 \neq 0$. Define

$$w = \left(\frac{1}{u\bar{u}}\right)\bar{u}$$

(we are scalar multiplying $\bar{u}$ by the real number $1/u\bar{u}$). Then

$$uw = \frac{1}{u\bar{u}} \cdot u\bar{u} = 1,$$

as desired. Here is an explicit formula.

$$\frac{1}{u} = \frac{1}{u\bar{u}} \cdot \bar{u} = \frac{1}{a^2 + b^2}(a - ib) = \frac{a}{a^2 + b^2} - i\frac{b}{a^2 + b^2}. \quad \bullet$$

*Example.* If $u = 2 + 5i$, then $u\bar{u} = 4 + 25 = 29$, and

$$\frac{1}{2 + 5i} = \tfrac{1}{29}(2 - 5i) = \tfrac{2}{29} - \tfrac{5}{29}i.$$

We check:

$$(2 + 5i)\left[\tfrac{2}{29} - \tfrac{5}{29}i\right] = \tfrac{4}{29} - \tfrac{25}{29}i^2 = \tfrac{4}{29} + \tfrac{25}{29} = \tfrac{29}{29} = 1.$$

*Derision.* Let $z = a + bi$ and $u = c + di$. If $u \neq 0$, define $z/u$ to be the product $z \cdot (1/u)$:

$$\frac{z}{u} = \frac{a + bi}{c + di} = a + bi \cdot \frac{1}{c + di}.$$

Therefore, if $z = a + ib$ and $u = c + id \neq 0$, then

$$\frac{z}{u} = \frac{z}{u} \cdot \frac{\bar{u}}{\bar{u}} = \frac{z\bar{u}}{u\bar{u}} = \frac{(a + ib)(c - id)}{c^2 + d^2}.$$

*Example*. Using the previous example,

$$\frac{3+7i}{2+5i} = (3+7i)\left[\tfrac{2}{29} - \tfrac{5}{29}i\right]$$

$$= \tfrac{1}{29}(3+7i)(2-5i)$$

$$= \tfrac{1}{29}(41-i) = \tfrac{41}{29} - \tfrac{1}{29}i.$$

**Theorem 4.10**. Every complex number $a + bi$ has a square root.

*Proof*. We must find real numbers $x$ and $y$ such that

$$(x+yi)^2 = a+bi.$$

That is, we need

$$a+bi = x^2 + 2xyi + (yi)^2$$

$$= x^2 - y^2 + 2xyi.$$

This gives two equations involving real numbers:

$$a = x^2 - y^2 \qquad\qquad (3)$$

and

$$b = 2xy. \qquad\qquad (4)$$

The case $b = 0$ is easy, for then $a + bi = a$ is a real number. If $a \geq 0$, then $\sqrt{a}$ is a real number; if $a < 0$, then $a = -p$, for some positive number $p$, and $\sqrt{a} = i\sqrt{p}$. Therefore, we may assume that $b \neq 0$, and, hence, that $x \neq 0$, by Eq. (4). Write $y = b/2x$. Substituting into Eq. (3), we obtain

$$x^2 - \left(\frac{b}{2x}\right)^2 = a.$$

Rewriting, we obtain

$$4x^4 - 4ax^2 - b^2 = 0,$$

a quadratic equation in $x^2$. The quadratic formula gives

$$x^2 = \tfrac{1}{8}\left(4a \pm \sqrt{16a^2 + 16b^2}\right)$$

$$= \tfrac{1}{2}\left(a \pm \sqrt{a^2 + b^2}\right).$$

Now $a^2 + b^2 > 0$, and so it has a real square root. Also,

$$a < \sqrt{a^2 + b^2}$$

because $b^2 > 0$, and so the root $\tfrac{1}{2}(a - \sqrt{a^2 + b^2})$ is negative. Since $x$ is supposed to be real, we may ignore this root; that is, we have

$$x^2 = \tfrac{1}{2}\left(a + \sqrt{a^2 + b^2}\right).$$

The right-hand side is positive, and so it has a real square root; that is, we can find $x$. But once we have $x$, we can find $y$, using Eq. (4). We have found both $x$ and $y$, and so we have found a square root of $a + bi$. •

There are two possible values for $x$, differing only in sign (and, hence, two possible values for $y$), as there should be, for if $x + yi$ is a square root of $a + bi$, then so is $-x - yi = -(x + yi)$.

As an example, let us compute $\sqrt{i}$. Here, $a = 0$ and $b = 1$, and Eqs. (3) and (4) give

$$\sqrt{i} = \pm\tfrac{1}{\sqrt{2}}(1 + i).$$

In the next section, we will give another way to find square roots of complex numbers.

**Theorem 4.11.** If $a$, $b$, and $c$ are complex numbers with $a \neq 0$, then the quadratic formula holds for $ax^2 + bx + c$.

*Proof.* The only possible obstruction is whether $b^2 - 4ac$ has a square root, and Theorem 4.10 shows that this is not a problem. •

In particular, the quadratic formula always holds whenever all the coefficients are real numbers, even when the discriminant $b^2 - 4ac$ is negative; in this last case, the roots are complex; indeed, they are conjugate (as you are asked to prove in Exercise 4.15).

The following consequence of the quadratic formula will be used in deriving the cubic formula.

**Corollary 4.12.** Given numbers $c$ and $d$, there exist numbers $g$ and $h$ with $g + h = c$ and $gh = d$.

**Proof.** If $d = 0$, choose $g = 0$ and $h = c$. If $d \neq 0$, define $g$ to be a root of $f(x) = x^2 - cx + d = 0$ [Theorem 4.11 shows that such a number $g$ exists], and define $h = d/g$ [by Exercise 4.1, 0 is not a root of $f(x)$ because $d \neq 0$)]. Clearly, $gh = d$ (so that $g \neq 0$), while $f(g) = 0$ gives

$$g + h = g + \frac{d}{g} = \frac{g^2 + d}{g} = \frac{cg}{g} = c.$$

Of course, $g$ and $h$ might be complex numbers, even when both $c$ and $d$ are real. •

Theorem 4.11 shows that every quadratic polynomial with complex coefficients has a (complex) root. This is a special case of the **Fundamental Theorem of Algebra** (first proved by C. F. Gauss (1777–1855) in 1799): For every $n \geq 1$, the polynomial

$$x^n + bx^{n-1} + cx^{n-2} + \cdots,$$

where the coefficients $b, c, \ldots$ are complex numbers, has a complex root.

We discuss some polynomials of higher degree in the next section. A simple polynomial is

$$x^3 - 1;$$

what are its roots? One root is obviously $x = 1$; what are the other two? Since

$$x^3 - 1 = (x - 1)(x^2 + x + 1),$$

we can find the other two roots with the quadratic formula:

$$\zeta = -\tfrac{1}{2} + i\tfrac{\sqrt{3}}{2} \quad \text{and} \quad \overline{\zeta} = -\tfrac{1}{2} - i\tfrac{\sqrt{3}}{2}$$

($\zeta$ is lower case Greek *zeta*). [As noted above, the roots must be complex conjugates.] The numbers $1$, $\zeta$, and $\overline{\zeta}$ are called the **cube roots of unity**. In Exercise 4.17, you are asked to check that $\overline{\zeta} = \zeta^2$.

Does every complex number $a + bi$ have a cube root? Is there a complex number $x + yi$ with $(x + yi)^3 = a + bi$? If we proceed as we did in the proof of Theorem 4.10, we obtain

$$a + bi = (x + yi)^3$$

$$= x^3 + 3x^2yi + 3xy^2i^2 + y^3i^3$$

$$= x^3 - 3xy^2 + (3x^2y - y^3)i.$$

This leads to intractable equations:

$$\left.\begin{array}{rcl} a & = & x^3 - 3xy^2 \\ b & = & 3x^2y - y^3 \end{array}\right\} \tag{5}$$

We will see how to find higher roots of complex numbers in the next section.

### *Exercises*

**4.8.** Rewrite each of the following complex numbers in the form $a + bi$.
  (i) $(3 + 4i)(2 - i)$.                               Answer: $10 + 5i$.
  (ii) $(1 + i)^2$.                                    Answer: $2i$.
  (iii) $\left(\frac{1}{\sqrt{2}} + i\frac{1}{\sqrt{2}}\right)^2$.                  Answer: $i$.
  (iv) $(3 + 4i)(3 - 4i)$.                             Answer: $25$.
  (v) $(3 + 4i)/(1 + i)$.                              Answer: $\frac{1}{2}(7 + i)$.

**4.9.** Prove the *cancellation law* for complex numbers: if $u$, $v$, and $z$ are complex numbers with $zu = zv$, and if $z \neq 0$, then $u = v$.

**4.10.** Show that if $u$ and $v$ are nonzero complex numbers, then their product $uv$ is also nonzero.

**4.11.** Prove that if $z$ is a complex number on the unit circle, then $1/z = \bar{z}$.

**4.12.** Let $z = a + bi$. Prove that if $z$ is a real number, then $z = \bar{z}$; conversely, if $z = \bar{z}$, then $z$ is a real number.

**4.13.** If $z = a + bi$, show that $(x - z)(x - \bar{z})$ is a quadratic polynomial having real coefficients.

**4.14.** If $z$ and $w$ are complex numbers, prove that

$$\overline{z + w} = \bar{z} + \bar{w} \quad \text{and} \quad \overline{zw} = \bar{z} \cdot \bar{w}$$

**4.15.** Let $f(x) = ax^2 + bx + c$, where the coefficients $a$, $b$, and $c$ are real. Prove that if the roots of $f(x)$ are not real numbers, then they are complex conjugates.

**4.16.** If $u$ is a root of a cubic polynomial $f(x)$ having real coefficients, then its conjugate $\bar{u}$ is also a root of $f(x)$.

**4.17.** If $\zeta$ is a cube root of unity, show that $\bar{\zeta} = \zeta^2$.

**4.18.** Find the square roots of $3 - 4i$.                Answer: $\pm(2 - i)$.

**4.19.** Find the roots of $x^2 + (2 + i)x + 2i = 0$.
                Answer: $-2, -i$. (Notice that the roots are not conjugate.)

**4.20.** Prove the binomial theorem for complex numbers: If $z$ and $w$ are complex numbers, then for all $n \geq 0$,

$$(z + w)^n = \sum_{r=0}^{n} \binom{n}{r} z^r w^{n-r}.$$

In particular, when $a$ and $b$ are real,

$$(a + ib)^n = \sum_{r=0}^{n} \binom{n}{r} a^r (ib)^{n-r} = \sum_{r=0}^{n} \binom{n}{r} a^r i^{n-r} b^{n-r}.$$

(Hint: The proofs of Lemma 1.18 and Corollary 1.21 work here as well.)

**4.21.** Prove that if $u$, $v$, and $w$ are roots of a cubic polynomial $x^3 + bx^2 + cx + d$, then $b = -(u + v + w)$ and $d = -uvw$. [Hint: $4x^3 + bx^2 + cx + d = (x - u)(x - v)(x - w)$.]

## DE MOIVRE'S THEOREM

> The science of mathematics presents the most brilliant example of how pure reason may successfully enlarge its domain without the aid of experience.
>
> *Immanuel Kant*

Just as there is a geometric way to view addition of complex numbers, so, too, is there a geometric way to view their multiplication. The real number $z\bar{z}$ has an obvious geometric meaning: if $z = a + ib$, then

$$\sqrt{z\bar{z}} = \sqrt{a^2 + b^2}$$

is the distance from $(a, b)$ to the origin. Recall that if $a$ is a real number, then its absolute value $|a|$ is the distance from $a$ to 0; hence, if $z = a + ib$, we write

$$|z| = \sqrt{z\bar{z}} = \sqrt{a^2 + b^2},$$

for it is the distance from $(a, b)$ to $(0, 0)$.

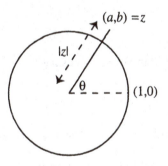

Figure 4.5

*Definition*. Call $|z|$ the *modulus* of $z$.

The unit circle can now be described as the collection of all complex numbers $x + iy$ of modulus 1; that is, the unit circle consists of all $x + iy = \cos\theta + i\sin\theta$ (see Figure 2.36).

***Theorem 4.13 (Polar Decomposition).*** Every complex number $z$ has a factorization

$$z = r(\cos \theta + i \sin \theta),$$

where $r = |z| \geq 0$ and $0 \leq \theta < 2\pi$.

*Proof.* If $z = 0$, then $|z| = 0$ and any choice of $\theta$ works. If $z = a + bi \neq 0$, then $|z| \neq 0$. Now $z/|z| = (a/|z|, b/|z|)$ has modulus 1, for $(a/|z|)^2 + (b/|z|)^2 = (a^2 + b^2)/|z|^2 = 1$. Therefore,

$$\frac{z}{|z|} = \cos \theta + i \sin \theta,$$

and so $z = |z|(z/|z|) = r(\cos \theta + i \sin \theta)$.   •

If $z = a + ib = r(\cos \theta + i \sin \theta)$, then $(r, \theta)$ are the ***polar coordinates*** of $z$; this is the reason Theorem 4.13 is called the polar decomposition of $z$.

The addition theorem of trigonometry has a lovely translation in the language of complex numbers.

***Theorem 4.14 (Addition Theorem).*** If $z = \cos \theta + i \sin \theta$ and $w = \cos \psi + i \sin \psi$, then

$$zw = \cos(\theta + \psi) + i \sin(\theta + \psi).$$

*Proof.*

$$zw = (\cos \theta + i \sin \theta)(\cos \psi + i \sin \psi)$$

$$= (\cos \theta \cos \psi - \sin \theta \sin \psi) + i(\sin \theta \cos \psi + \cos \theta \sin \psi).$$

The trigonometric addition formulas, Theorem 2.18, show that

$$zw = \cos(\theta + \psi) + i \sin(\theta + \psi).$$   •

It follows that the product of two complex numbers on the unit circle also lies on the unit circle.

***Example.*** Let us call $(a, b, c)$ a ***semi-Pythagorean triple*** if $a$, $b$ and $c$ are (possibly negative) integers, not all 0, for which $a^2 + b^2 = c^2$. Some examples of semi-Pythagorean triples are $(3, -4, 5)$, $(4, -3, -5)$, and $(3, 0, 3)$, as well as all Pythagorean triples. It is easy to see that if $(a, b, c)$ is a semi-Pythagorean triple with $0 < |a| < |b|$, then $(|a|, |b|, |c|)$ is a Pythagorean triple.

Each semi-Pythagorean triple $(a, b, c)$ determines a rational point on the unit circle, namely, $\frac{a}{c} + i\frac{b}{c}$. Conversely, every rational point on the unit circle of the form $\frac{a}{c} + i\frac{b}{c}$ determines a semi-Pythagorean triple, namely, $(a, b, c)$. Therefore, semi-Pythagorean triples correspond to rational points on the unit circle, whereas Pythagorean triples correspond to rational points on the unit circle in the first quadrant between $(\frac{1}{\sqrt{2}}, \frac{1}{\sqrt{2}})$ and $(0, 1)$. We noted above that the product of two points on the unit circle is another such, and it is easy to see that the product of two rational points on the unit circle is rational as well. The formula for complex multiplication:

$$\left(\frac{a}{c} + i\frac{b}{c}\right)\left(\frac{r}{t} + i\frac{s}{t}\right) = \frac{ar - bs}{ct} + i\frac{as + br}{ct},$$

suggests a definition of "multiplication" of semi-Pythagorean triples:

$$(a, b, c) * (r, s, t) = (ar - bs, as + br, ct),$$

and the "product" of semi-Pythagorean triples is again a semi-Pythagorean triple. For example, the "square" $(3, 4, 5) * (3, 4, 5) = (-7, 24, 25)$, and this gives the Pythagorean triple $(7, 24, 25)$. More generally,

$$(a, b, c) * (a, b, c) = (a^2 - b^2, 2ab, a^2 + b^2),$$

a formula reminiscent of the classification of Pythagorean triples.

In 1707, A. De Moivre (1667–1754) proved the following elegant result.

**Theorem 4.15 (De Moivre).** For every real number $\theta$ and every positive integer $n$,

$$\cos n\theta + i\sin n\theta = (\cos\theta + i\sin\theta)^n.$$

*Remark.* In view of Exercise 4.20, $\cos n\theta$ and $\sin n\theta$ can be computed in terms of $\cos\theta$ and $\sin\theta$ just by using the binomial theorem.

*Proof.* We prove De Moivre's theorem by induction on $n \geq 1$. The base step $n = 1$ is obviously true. For the inductive step,

$$(\cos\theta + i\sin\theta)^{n+1} = (\cos\theta + i\sin\theta)^n(\cos\theta + i\sin\theta)$$

$$= (\cos n\theta + i\sin n\theta)(\cos\theta + i\sin\theta)$$

(inductive hypothesis)

$$= \cos(n\theta + \theta) + i\sin(n\theta + \theta)$$

(addition formula)

$$= \cos([n+1]\theta) + i\sin([n+1]\theta). \quad \bullet$$

**Example.** Find the value of $(\cos 3° + i\sin 3°)^{40}$.

By De Moivre's theorem,

$$(\cos 3° + i\sin 3°)^{40} = \cos 120° + i\sin 120° = -\tfrac{1}{2} + i\tfrac{\sqrt{3}}{2}.$$

**Corollary 4.16.**

(i)   $\cos 2\theta = \cos^2\theta - \sin^2\theta = 2\cos^2\theta - 1$
      $\sin 2\theta = 2\sin\theta\cos\theta.$

(ii)  $\cos 3\theta = \cos^3\theta - 3\cos\theta\sin^2\theta = 4\cos^3\theta - 3\cos\theta$
      $\sin 3\theta = 3\cos^2\theta\sin\theta - \sin^3\theta = 3\sin\theta - 4\sin^3\theta.$

*Proof.* (i)

$$\cos 2\theta + i\sin 2\theta = (\cos\theta + i\sin\theta)^2$$
$$= \cos^2\theta + 2i\sin\theta\cos\theta + i^2\sin^2\theta$$
$$= \cos^2\theta - \sin^2\theta + i(2\sin\theta\cos\theta).$$

Equating real and imaginary parts gives both double angle formulas.

(ii) De Moivre's theorem gives

$$\cos 3\theta + i \sin 3\theta = (\cos \theta + i \sin \theta)^3$$
$$= \cos^3 \theta + 3i \cos^2 \theta \sin \theta + 3i^2 \cos \theta \sin^2 \theta + i^3 \sin^3 \theta$$
$$= \cos^3 \theta - 3 \cos \theta \sin^2 \theta + i(3 \cos^2 \theta \sin \theta - \sin^3 \theta).$$

Equality of the real parts gives $\cos 3\theta = \cos^3 \theta - 3 \cos \theta \sin^2 \theta$; the second formula for $\cos 3\theta$ follows by replacing $\sin^2 \theta$ by $1 - \cos^2 \theta$. Equality of the imaginary parts gives $\sin 3\theta = 3 \cos^2 \theta \sin \theta - \sin^3 \theta = 3 \sin \theta - 4 \sin^3 \theta$; the second formula arises by replacing $\cos^2 \theta$ by $1 - \sin^2 \theta$.   •

Compare the first formulas for $\cos 3\theta$ and $\sin 3\theta$ with Eqs. (5) on page 169.

We are going to present a beautiful formula discovered by Euler, but first we must recall some power series formulas to see how it arises. For every real number $x$,

$$e^x = 1 + x + \frac{x^2}{2!} + \cdots + \frac{x^n}{n!} + \cdots ,$$

$$\cos x = 1 - \frac{x^2}{2!} + \frac{x^4}{4!} - \cdots + \frac{(-1)^n x^{2n}}{(2n)!} + \cdots ,$$

and

$$\sin x = x - \frac{x^3}{3!} + \frac{x^5}{5!} - \cdots + \frac{(-1)^{n-1} x^{2n+1}}{(2n+1)!} + \cdots .$$

**Theorem 4.17 (Euler).** For all real numbers $x$,

$$e^{ix} = \cos x + i \sin x.$$

*Sketch of proof.* As $n$ varies over $0, 1, 2, 3, 4, 5, \ldots$, the powers of $i$ repeat every four steps: that is, for $n \geq 0$, $i^n$ takes values

$$1, i, -1, -i, 1, i, -1, -i, 1, i, -1, -i, \ldots ;$$

Thus, the even powers of $i$ are all real, whereas the odd powers all involve $i$. It follows, for every real number $x$, that $(ix)^n$ takes values

$$1, ix, -x^2, -ix^3, x^4, ix^5, -x^6, -ix^7, x^8, ix^9, -x^{10}, -ix^{11}, \ldots .$$

One can define convergence of power series $\sum_{n=0}^{\infty} c_n z^n$ for $z$ and $c_n$ complex numbers, and one can show that the series

$$1 + z + \frac{z^2}{2!} + \cdots + \frac{z^n}{n!} + \cdots$$

converges for every complex number $z$. The **complex exponential** $e^z$ is defined to be the sum of this series. In particular, the series for $e^{ix}$ converges for all real numbers $x$, and

$$e^{ix} = 1 + ix + \frac{(ix)^2}{2!} + \cdots + \frac{(ix)^n}{n!} + \cdots .$$

The even powers of $ix$ do not involve $i$, whereas the odd powers do. Collecting terms, one has $e^{ix} =$ even terms $+$ odd terms. But

$$\text{even terms} = 1 + \frac{(ix)^2}{2!} + \frac{(ix)^4}{4!} + \cdots$$

$$= 1 - \frac{x^2}{2!} + \frac{x^4}{4!} - \cdots$$

and

$$\text{odd terms} = ix + \frac{(ix)^3}{3!} + \frac{(ix)^5}{5!} + \cdots .$$

$$= i(x - \frac{x^3}{3!} + \frac{x^5}{5!} - \cdots).$$

Therefore, $e^{ix} = \cos x + i \sin x$.   •

It is said that Euler was especially pleased with the equation

$$e^{i\pi} = -1;$$

indeed, this formula is inscribed on his tombstone.

As a consequence of Euler's theorem, the polar decomposition can be rewritten in exponential form: every complex number $z$ has a factorization

$$z = re^{i\theta},$$

where $r \geq 0$ and $0 \leq \theta < 2\pi$.

We have chosen to denote $\sum_{n=0}^{\infty} \frac{(ix)^n}{n!}$ by $e^{ix}$, but we cannot assert, merely as a consequence of our notation, that the law of exponents, $e^{ix}e^{iy} = e^{i(x+y)}$, is valid. However, this is precisely what Theorem 4.14 says once it is translated into exponential notation.

***Theorem 4.18 = Theorem 4.14 (Addition Theorem)*** For all real numbers $x$ and $y$,

$$e^{ix}e^{iy} = e^{i(x+y)}.$$

*Proof.* According to Theorem 4.14, $e^{ix}e^{iy} = (\cos x + i \sin x)(\cos y + i \sin y)$ $= \cos(x + y) + i \sin(x + y) = e^{i(x+y)}$.  •

It is easier to remember the trigonometric addition formulas in complex form. The addition formula gives almost all the standard identities of trigonometry! For example, Corollary 4.16 gives the double and triple angle formulas. One can now begin to believe that complex numbers are good for something.

We may also translate De Moivre's theorem into exponential notation.

***Theorem 4.19 = Theorem 4.15 (De Moivre).*** For every real number $x$ and every integer $n \geq 1$,

$$(e^{ix})^n = e^{inx}.$$

*Proof.* According to Theorem 4.15, $(e^{ix})^n = (\cos x + i \sin x)^n = \cos nx + i \sin nx = e^{inx}$.  •

Here is a geometric interpretation of complex multiplication. The polar decomposition $z = r(\cos \theta + i \sin \theta)$ of a complex number can now be written in exponential form, as $z = re^{i\theta}$, where $(r, \theta)$ are the polar coordinates of $z$. If $w = se^{i\psi}$, then Theorem 4.18 gives

$$zw = (re^{i\theta})(se^{i\psi}) = rse^{i(\theta+\psi)}.$$

Thus, if the polar coordinates of $z$ and $w$ are $(r, \theta)$ and $(s, \psi)$, respectively, then the polar coordinates of $zw$ are $(rs, \theta + \psi)$. One consequence of this equation is that $|zw| = rs$; that is,

$$|zw| = |z| \cdot |w|.$$

**Definition.** If $n \geq 1$ is an integer, then an ***nth root of unity*** is a complex number $\zeta$ with $\zeta^n = 1$.

The geometric interpretation of complex multiplication is particularly interesting when $z$ and $u$ lie on the unit circle, so that $r = 1 = s$. Given a positive integer $n$, let $\theta = 2\pi/n$ (we have switched to radian measure), and let $\zeta = e^{i\theta} = \cos\theta + i\sin\theta$. By De Moivre's theorem,

$$\zeta^n = (e^{2\pi i/n})^n = e^{2\pi i} = \cos 2\pi + i\sin 2\pi = 1,$$

so that $\zeta$ is an $n$th root of unity; indeed, for $k = 1, 2, 3, \ldots$, the powers $\zeta^k$ are also $n$th roots of unity, for

$$[\zeta^k]^n = [(e^{2\pi i/n})^k]^n = (e^{2\pi i})^k = 1^k = 1.$$

Thus, if the polar coordinates of $\zeta$ are $(1, \theta)$, then the polar coordinates of $\zeta^2$ are $(1, 2\theta)$, the polar coordinates of $\zeta^3$ are $(1, 3\theta)$, $\ldots$, the polar coordinates of $\zeta^{n-1}$ are $(1, [n-1]\theta)$; if $\zeta$ is an $n$th root of unity, then $\zeta^n = (1, 0)$. In this way, we see that the $n$th roots of unity are equally spaced around the unit circle. Figure 4.6 shows the 8th roots of unity.

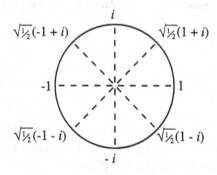

Figure 4.6

**Corollary 4.20.** Every $n$th root of unity is equal to

$$e^{2\pi i k/n} = \cos(2\pi k/n) + i\sin(2\pi k/n),$$

for some $k$ with $0 \leq k \leq n - 1$.

*Sketch of proof.* It is proved in an Abstract Algebra course (e.g., see Rotman, p. 176) that a polynomial of degree $n$ with complex coefficients has at most $n$ complex roots. In particular, $x^n - 1$ has at most $n$ roots; that is, there are at most $n$ $n$th roots of unity. As we saw above, $(\zeta^k)^n = 1$ for all $k$, and so we have displayed $n$ distinct $n$th roots of unity, namely,

$$1, \zeta, \zeta^2, \ldots, \zeta^{n-1};$$

as there are no more than $n$ such roots of unity, this must be all of them.    •

Earlier, we found the cube root of unity $\zeta = -\frac{1}{2} + i\frac{\sqrt{3}}{2}$ as a root of $x^3 - 1 = (x - 1)(x^2 + x + 1)$; with De Moivre's theorem, we find $\zeta$ as

$$e^{2\pi i/3} = \cos(\tfrac{2\pi}{3}) + i \sin(\tfrac{2\pi}{3}) = -\tfrac{1}{2} + i\tfrac{\sqrt{3}}{2}.$$

What are the $n$th roots of a nonzero number $g$? If $n = 2$, then the square roots of unity are 1 and $-1$, and so $g$ has two square roots: $\sqrt{g}$ and $-\sqrt{g}$. More generally, if $\gamma$ is an $n$th root of $g$, then there are $n$ $n$th roots of $g$, namely, $\gamma, \zeta\gamma, \zeta^2\gamma, \ldots, \zeta^{n-1}\gamma$, where $\zeta = e^{2\pi i/n}$.

De Moivre's theorem gives a way to find square roots of complex numbers other than that given in the proof of Theorem 4.10. Since $(e^{i\theta/2})^2 = e^{i\theta}$, we see that a square root of $e^{i\theta}$ is $e^{i\theta/2}$. Using the polar decomposition, we see that if $z = re^{i\theta} = r(\cos\theta + i\sin\theta)$, then

$$\sqrt{z} = \sqrt{r}e^{i\theta/2} = \sqrt{r}[\cos(\theta/2) + i\sin(\theta/2)].$$

For example, let us now find $\sqrt{5 + 12i}$. Here, $r^2 = 169$ and $r = 13$. A computer or a trigonometric table gives $\frac{5}{13} \approx \cos(67.3°)$ and $\frac{12}{13} \approx \sin(67.3°)$, so that

$$5 + 12i \approx 13[\cos(67.3°) + i\sin(67.3°)].$$

Hence,

$$\sqrt{5 + 12i} \approx \sqrt{13}[\cos(33.6°) + i\sin(33.6°)]$$

$$\approx 3.61(.832 + .555i) \approx 3 + 2i.$$

In fact, one can see that $\sqrt{5 + 12i} = \pm(3 + 2i)$. In these calculations, it is a matter of taste whether one uses radians or degrees.

One of my friends was in his office, seated at his desk, with the door open. A student he didn't know saw the open door and asked whether she might be able to get some help with computing $e^{it}$. My friend began explaining De Moivre's theorem, but it soon became clear that the student was not understanding him. "Are you comfortable with complex numbers?" The student was stunned. "Isn't $i$ the square root of $-1$?" my friend asked. "No," the student replied, "$i$ stands for interest."

Why is a root so called? Just as the Greeks called the bottom side of a triangle its base (as in the area formula $\frac{1}{2}$ height $\times$ base), they also called the bottom side of a square its base. A natural question for the Greeks was: Given a square of area $A$, what is the length of its side; of course, the answer is $\sqrt{A}$. Were we inventing a word for $\sqrt{A}$, we might have called it the *base* of $A$ or the *side* of $A$. Similarly, were we seeking a term for the analogous three-dimensional question: What is the length of a side of a cube of volume $V$, we might have called $\sqrt[3]{V}$ the *cube* base of $V$ and $\sqrt{A}$ the *square base* of $A$. Why, then, do we call these numbers cube *root* and square *root*? What has any of this to do with botany?

Because tracing the etymology of words is not a simple matter, we only suggest the following explanation. Until about the fourth and fifth centuries, most mathematics was written in Greek but, by the fifth century, India had become a center of mathematics and important mathematical texts were also written in Sanskrit. The Sanskrit term for square root is *pada*. Both Sanskrit and Greek are Indo-European languages, and the Sanskrit word *pada* is a cognate of the Greek word *podos*; both mean *base* in the sense of the foot of a pillar or, as above, the bottom of a square. In both languages, however, there is a secondary meaning: the root of a plant. In translating from Sanskrit, Arab mathematicians chose the secondary meaning, perhaps in error (Arabic is not an Indo-European language), perhaps for some reason unknown to us. For example, the influential book by al-Khwarizmi, *Al-jabr w'al muqabala*, which appeared in the year 830, used the Arabic word *jidhr*, meaning root of a plant. (The word *algebra* is a European version of the first word in the title of this book; the author's name has also come into the English language, as the word *algorithm*.) This mistranslation has since been handed down through the centuries; the term *jidhr* became standard in mathematical writing in Arabic, and European translations from Arabic into Latin used the word *radix* (meaning root, as in radish or radical). The notation $r2$ for $\sqrt{2}$ occurs in European writings from about the twelfth century (but the square root symbol did not arise from the letter $r$; it evolved from an old dot notation). However, there was a competing notation in use at the same time, for some scholars who translated directly from the Greek denoted $\sqrt{2}$ by $\ell 2$, where $\ell$ abbreviates the Latin word *latus*, meaning side. Finally, with the invention of logarithms in the 1500's, $r$ won out over $\ell$, for the notation $\ell 2$ was then commonly used to denote log 2. Generalizing this usage from square root and cube root to a root of any polynomial, not merely $x^2 - A$ or $x^3 - V$, is a natural enough step. Thus, as pleasant as it would be, roots of polynomials seem to have no botanical connection.

### *Exercises*

**4.22.** (i) Find $\sqrt{8 + 15i}$. (Hint: Write $z = 8 + 15i$ in polar form: $z = 17[\cos(62°) + i \sin(62°)]$, where $(17, 62°)$ are the polar coordinates of the point $(8, 15)$. (Hint: $\sin(31°) \approx .483046$ and $\cos(31°) \approx .885584$.)
(ii) Find $\sqrt[4]{8 + 15i}$. (Hint: $\sin(15.5°) \approx .249404$ and $\cos(15.5°) \approx .968399$.)

**4.23.** Find an 8th root of $9 - 7i$.                    Answer: $1.034 + .876i$.

**4.24.** Show that $\cos(1.25664) + i \sin(1.25664)$ is approximately a fifth root of unity ($1.25664$ radians $\approx 2\pi/5$).

**4.25.** Prove that if $\zeta$ is a complex cube root of unity, then

$$\zeta(1 - \zeta^2)(1 - \zeta)^2 = 3i\sqrt{3}.$$

[Hint: $\zeta(1 - \zeta^2)(1 - \zeta)^2 = \zeta(1 - \zeta^3)(1 + \zeta) = (1 - \zeta)^3(\zeta + \zeta^2)$. Use the following facts to simplify this expression: $\zeta^2 + \zeta + 1 = 0$, $\zeta^2 = \bar{\zeta}$, and $\zeta = -\frac{1}{2} \pm i\frac{\sqrt{3}}{2}$.]

**4.26.** For every positive integer $n$, show that there is a polynomial $f_n(x)$ of degree $n$ and with integer coefficients so that $\cos n\theta = f_n(\cos \theta)$. [Hint: Use De Moivre's theorem.]

**4.27.** (i) Prove that $2 \cos \theta = e^{i\theta} + e^{-i\theta}$.
(ii) Use De Moivre's theorem to give a new proof of Exercise 2.51:

$$2 \cos(n + 1)\theta = (2 \cos \theta)(2 \cos n\theta) - 2 \cos(n - 1)\theta.$$

**4.28.** (i) It can be proved that if $z = a + ib$, then

$$e^z = e^a e^{ib} = e^a(\cos b + i \sin b)$$

(recall that $e^z$ is defined as the sum of a certain series). In contrast to real exponentiation, show that $e^z$ can be a negative real number.

(ii) If $w = e^z$, where $z$ is a complex number, then define $\log(w) = z$. In contrast to real logarithms, show that $-1$ has a logarithm; indeed, show that $-1$ has infinitely many logarithms.

**4.29.** For any complex number $z$, define

$$\cos z = \tfrac{1}{2}(e^{iz} + e^{-iz}).$$

[Recall Exercise 4.27(i): If $x$ is a real number, then $\cos x = \tfrac{1}{2}(e^{ix} + e^{-ix})$.] In contrast to real cosines, show that $\cos z > 1$ is possible. (Hint: Show that if $b$ is real, then $\cos(ib) = \cosh(b)$.)

**4.30.** (i) Write $1/(e^{i\theta} - 1)$ in the form $a + ib$.

(ii) Prove, for all $n \geq 1$ and for all $\theta$, that

$$1 + e^{i\theta} + e^{i2\theta} + \cdots + e^{in\theta} = \frac{e^{i(n+1)\theta} - 1}{e^{i\theta} - 1}.$$

(iii) Prove the identity

$$(2 - 2\cos\theta)(\sin\theta + \sin 2\theta + \sin 3\theta)$$
$$= -(\cos 4\theta - 1)\sin\theta + \sin 4\theta(\cos\theta - 1).$$

## CUBICS AND QUARTICS

Quando che'l cubo con le cose appresso
Se agguaglia a qualche numero discreto:
Trovan dui altri, differenti in esso.
Dapoi terrai, questo per consueto,
Che'l lor produtto, sempre sia eguale
Al terzo cubo delle cose neto;
El residuo poi suo generale,
Delli lor lati cubi, bene sottratti
Varra la tua cose principale.... .[16]

*Niccoló Fontana (Tartaglia)*

Revolutionary events were changing the western world at the beginning of the sixteenth century: the printing press had just been invented; trade with Asia and Africa was flourishing; Columbus had just discovered the New World; Martin Luther was challenging papal authority. The Reformation and the Renaissance were beginning.

The Italian peninsula was not one country but a collection of city states with many wealthy and cosmopolitan traders. And there were public mathematics contests sponsored by the dukes of the cities. In particular, one of the problems often set involved finding roots of a given cubic

$$x^3 + bx^2 + cx + d,$$

where $b$, $c$, and $d$ were real numbers, usually integers.

We must remark that modern notation did not exist in the early 1500's, and so the feat of finding roots involved not only mathematical ingenuity but also an ability to surmount linguistic obstacles. Designating variables by letters was invented by F. Viète (1540–1603) in 1591, who used consonants to denote constants and vowels to denote variables (the modern notation using the letters $a$, $b$, $c$, ... to denote constants and letters $x$, $y$, $z$ at the end of the alphabet to denote variables was introduced by Descartes in 1637). The exponential notation $x^2$, $x^3$, $x^4$, ... was essentially introduced by J. Hume in 1636 (he used $2^{ii}$, $2^{iii}$, $2^{iv}$, ...). The symbols $+$, $-$, and $\sqrt{\ }$, as well as the symbol / for division (as in 4/5), were introduced by J. Widman in 1486 (the symbol $\div$ for division was introduced by J. H. Rahn in 1659). The symbol $\times$ for

---

[16]A rough translation of this verse is given below in the quotation from the book of Tignol.

multiplication was introduced by W. Oughtred in 1631. The symbol = was introduced by R. Recorde in 1557 in his *Whetstone of Wit*:

> "And to avoide the tediouse repetition of these woordes: is equal to: I will lette as I doe often in woorke use, a paire of paralleles, or gemowe lines of one lengthe, thus: = , because noe 2 thynges, can be moare equalle."

(*Gemowe* is an obsolete word, meaning *twin* or, in this case, *parallel*.) These symbols were not adopted at once, and often there were competing notations. Only in the seventeenth century, with the publication of Descartes's book *La Géométrie* in 1637, did most modern notation for arithmetic become universal in Europe.

Let us return to cubic polynomials. The lack of a good notation was a great handicap. For example, the cubic equation $x^3 + 2x^2 - 4x + 1 = 0$ would be given, roughly, as follows:

> Take the cube of a thing, add to it twice its square, add 1, and this must all be equal to 4 times the thing.

Complicating matters even more, negative numbers were not accepted; thus, an equation of the form $x^3 - 2x^2 - 4x + 1 = 0$ would be given only in the form $x^3 + 1 = 2x^2 + 4x$. Thus, there were many types of cubic equations, depending (in our notation) on whether coefficients were positive, negative, or zero.

About 1515, Scipione del Ferro of Bologna discovered a formula for the roots of several types of a cubic. Given the competitive context, it was natural for him to keep his method secret. Before his death in 1526, Scipione shared his result with some of his students.

The following history is a quotation from the excellent account in Tignol's book, *Galois' Theory of Equations*.

> "In 1535, Niccolò Fontana (c.1500–1557), nicknamed "Tartaglia" ("Stammerer"), from Brescia, who had dealt with some very particular cases of cubic equations, was challenged to a problem-solving contest by Antonio Maria Fior, a former pupil of Scipione del Ferro. When he heard that Fior had received the solution of cubic equations from his master, Tartaglia threw all his energy and skill into the struggle. He succeeded in finding the solution just in time to inflict upon Fior a humiliating defeat.

The news that Tartaglia had found the solution of cubic equations reached Giralamo Cardano (1501–1576), a very versatile scientist, who wrote a number of books on a wide variety of subjects, including medicine, astrology, astronomy, philosophy, and mathematics. Cardano then asked Tartaglia to give him his solution, so that he could include it in a treatise on arithmetic, but Tartaglia flatly refused, since he was himself planning to write a book on this topic. It turns out that Tartaglia later changed his mind, at least partially, since in 1539 he handed on to Cardano the solution of $x^3 + qx = r$, $x^3 = qx + r$, and a very brief indication of $x^3 + r = qx$ in verses.

The excerpt of Tartaglia's poem which is quoted at the beginning of this section gives the formula for $x^3 + qx = r$. The equation is indicated in the first two verses: the cube and the things equal to a number. *Cosa* (= thing) is the word for the unknown. To express the fact that the unknown is multiplied by a coefficient, Tartaglia simply uses the plural form: *le cose*. He then gives the following procedure: find two numbers which differ by the given number and such that their product is equal to the cube of the third of the number of things. Then the difference between the cube roots of these numbers is the unknown.

With modern notations, we would write that, to find the solution of $x^3 + qx = r$, we only need find $t$ and $u$ such that

$$\left. \begin{array}{l} t - u = r \\[2mm] tu = (q/3)^3 \end{array} \right\} \tag{6}$$

then

$$x = \sqrt[3]{t} - \sqrt[3]{u}.$$

The values of $t$ and $u$ are easily found. (Eqs. (6) lead to the quadratic equation $t^2 - rt - (\frac{1}{3}q)^3 = 0$) ... However, the poem does not provide any justification for the formula. Of course, it "suffices" to check that the value of $x$ given above satisfies the equation $x^3 + qx = r$ (that is, substitute the given value for $x$ into the cubic and see whether one obtains 0), but this was far from obvious to a sixteenth century mathematician .... The major difficulty was to figure out that

$$(a - b)^3 = a^3 - 3a^2b + 3ab^2 - b^3,$$

a formula which could be properly proved only by a dissection of a cube in three-dimensional space.

Having received Tartaglia's poem, Cardano set to work. Not only did he find justifications for the formulas, but he also solved all the other types of cubics. He then published his results, giving due credit to Tartaglia and to del Ferro, in the epoch-making book *Ars Magna, sive de regulis algebraicis* (The Great Art, or the Rules of Algebra)."

The proof of the cubic formula will involve replacing a given cubic polynomial by a simpler one.

**Definition.** A cubic $x^3 + qx + r$ having no $x^2$ term is called a *reduced* cubic.

**Lemma 4.21.** The substitution $X = x - \frac{1}{3}b$ changes a cubic

$$f(X) = X^3 + bX^2 + cX + d$$

into a reduced cubic

$$\tilde{f}(x) = f(x - \tfrac{1}{3}b) = x^3 + qx + r;$$

moreover, if $u$ is a root of $\tilde{f}(x)$, then $u - \frac{1}{3}b$ is a root of $f(X)$.

*Proof.* The substitution $X = x - \frac{1}{3}b$ gives

$$f(X) = f(x - \tfrac{1}{3}b)$$

$$= (x - \tfrac{1}{3}b)^3 + b(x - \tfrac{1}{3}b)^2 + c(x - \tfrac{1}{3}b) + d$$

$$= x^3 - bx^2 + \tfrac{1}{3}b^2 x - \tfrac{1}{27}b^3 + bx^2 - \tfrac{2}{3}b^2 x + \tfrac{1}{9}b^3 + cx - \tfrac{1}{3}bc + d.$$

As the $x^2$ terms cancel, this new polynomial $\tilde{f}(x) = f(x - \frac{1}{3}b)$ is reduced: $\tilde{f}(x) = x^3 + qx + r$, where the coefficients $q$ and $r$ can be found from the last equation by collecting terms.

Finally, if $u$ is a root of $\tilde{f}(x)$, then $0 = \tilde{f}(u) = f(u - \frac{1}{3}b)$; that is, $u - \frac{1}{3}b$ is a root of $f(X)$. •

The lemma reduces the problem of finding the roots of a cubic polynomial to the problem of finding the roots of a cubic having no $x^2$ term. The "trick" in solving the cubic is to write a root $u$ of $x^3 + qx + r$ as

$$u = g + h,$$

and then to find $g$ and $h$. Hence,

$$0 = (g + h)^3 + q(g + h) + r.$$

Note that

$$(g + h)^3 = g^3 + 3g^2h + 3gh^2 + h^3$$

$$= g^3 + h^3 + 3(g^2h + gh^2)$$

$$= g^3 + h^3 + 3gh(g + h)$$

$$= g^3 + h^3 + 3ghu.$$

Substituting $u = g + h$ into $0 = u^3 + qu + r$ now gives

$$0 = g^3 + h^3 + 3ghu + qu + r$$
$$= g^3 + h^3 + u(3gh + q) + r. \tag{7}$$

By Corollary 4.12, we may impose a second condition

$$gh = -q/3 \tag{8}$$

(we have already set $g + h = u$); thus, the $u$ term in Eq. (7) goes away, leaving

$$g^3 + h^3 = -r. \tag{9}$$

Cubing each side of Eq. (8) gives

$$g^3h^3 = -q^3/27. \tag{10}$$

Eqs. (9) and (10) in the two unknowns $g^3$ and $h^3$ can be solved, as in Corollary 4.12.

Notice that $g = 0$ precisely when $q = 0$. In this case, the original cubic is just $f(x) = x^3 + r$ whose roots are just the three cube roots of $-r$.

We may, therefore, assume that $g \neq 0$. Substituting $h^3 = -q^3/27g^3$ into Eq. (9) gives

$$g^3 - \frac{q^3}{27g^3} = -r,$$

which may be rewritten as

$$g^6 + rg^3 - \tfrac{1}{27}q^3 = 0,$$

a quadratic in $g^3$. The quadratic formula gives

$$g^3 = \tfrac{1}{2}\left(-r + \sqrt{R}\right),$$

where $R = r^2 + \tfrac{4}{27}q^3$, and taking a cube root gives $g$. Using Eq. (8), we can now solve for $h = -q/3g$, and $u = g + h$. [The number $z = \tfrac{1}{2}(-r + \sqrt{R})$ might be complex; the easiest way to find a cube root of $z$ is to write it in polar form and use De Moivre's theorem.]

*Remark.* One can give a formula for $h$ in terms of the coefficients of $f(x)$. Eq. (9) gives

$$
\begin{aligned}
h^3 &= -r - g^3 \\[2mm]
&= -r - \tfrac{1}{2}\left(-r + \sqrt{R}\right) \\[2mm]
&= \tfrac{1}{2}\left(-r - \sqrt{R}\right).
\end{aligned}
$$

We have found one root $u = g + h$ of a cubic; what are the other two roots? Exercise 4.33 below says that if $u$ is a root of a polynomial $f(x)$, then $f(x) = (x - u)d(x)$ for some polynomial $d(x)$. After finding one root $u = g + h$, divide $x^3 + qx + r$ by $x - u$, and use the quadratic formula on the quadratic quotient $d(x)$ to find the other two roots [if $\beta$ is a root of $d(x)$, then $\beta$ is a root of $f(x)$, for if $d(\beta) = 0$, then $f(\beta) = (\beta - u)d(\beta) = 0$]. (This is the way we first found the complex cube roots of unity.)

Here is an explicit formula for the other two roots of $f(x)$ (instead of the method just given for finding them). There are three cube roots of unity, namely, $1, \zeta = -\tfrac{1}{2} + i\tfrac{\sqrt{3}}{2}$, and $\zeta^2 = -\tfrac{1}{2} - i\tfrac{\sqrt{3}}{2}$, and so there are three cube roots of $g^3$, namely, $g, \zeta g$ and $\zeta^2 g$. We have already chosen $g$ and its mate $h = -q/3g$, and these give the root $u = g + h$. If we now choose $\zeta g$ instead of $g$, then its mate is $-q/3\zeta g = (1/\zeta)h = \zeta^2 h$, and these give the root $\zeta g + \zeta^2 h$. Finally, if we choose $\zeta^2 g$ instead of $g$, then its mate is $-q/3\zeta^2 g = (1/\zeta^2)h = \zeta h$, and these give the root $\zeta^2 g + \zeta h$. We have proved the cubic formula (also called *Cardano's formula*).

***Theorem 4.22 (Cubic Formula).*** The roots of $x^3 + qx + r$ are

$$g + h, \quad \zeta g + \zeta^2 h, \quad \text{and} \quad \zeta^2 g + \zeta h,$$

where $g = \sqrt[3]{\frac{1}{2}(-r + \sqrt{R})}, h = -q/3g, R = r^2 + \frac{4}{27}q^3$, and $\zeta = -\frac{1}{2} + i\frac{\sqrt{3}}{2}$ is a cube root of unity.

***Remark.*** One can rewrite $R = 4\left[(\frac{r}{2})^2 + (\frac{q}{3})^3\right]$, and the rewritten formulas for $g^3$ and $h^3$ are now reminiscent of the quadratic formula:

$$g^3 = \tfrac{1}{2}(-r + \sqrt{R}) = -\tfrac{1}{2}r + \sqrt{(\tfrac{r}{2})^2 + (\tfrac{q}{3})^3},$$

and

$$h^3 = \tfrac{1}{2}(-r - \sqrt{R}) = -\tfrac{1}{2}r - \sqrt{(\tfrac{r}{2})^2 + (\tfrac{q}{3})^3}.$$

***Good Example.*** Find the roots of

$$x^3 - 15x - 126.$$

Because there is no $x^2$ term, the polynomial is already reduced (otherwise, one would first reduce it as in Lemma 4.21). Here, $q = -15, r = -126, R = (-126)^2 + \frac{4}{27}(-15)^3 = 15376$, and $\sqrt{R} = 124$. Hence,

$$g^3 = \tfrac{1}{2}[-(-126) + 124] = 125 \text{ and } g = 5,$$

so that

$$h = -q/(3g) = 15/15 = 1.$$

The roots are thus: $g + h = 6$, $\zeta g + \zeta^2 h = -3 + 2i\sqrt{3}$, and $\zeta^2 g + \zeta h = -3 - 2i\sqrt{3}$. (It is proved in calculus that every cubic polynomial having real coefficients has at least one real root; one can also prove this using the cubic formula; see Exercise 4.34.)

Alternatively, having found $u = 6$ to be a root, long division gives

$$x^3 - 15x - 126 = (x - 6)(x^2 + 6x + 21),$$

and the quadratic formula gives $-3 + 2i\sqrt{3}$ and $-3 - 2i\sqrt{3}$ as the roots of the quadratic factor.

**Bad Example**. Let us try the cubic formula on the polynomial

$$x^3 - 7x + 6 = (x - 1)(x - 2)(x + 3)$$

whose roots are, obviously, 1, 2, and $-3$. There is no $x^2$ term, $q = -7, r = 6$, and $R = r^2 + \frac{4}{27}q^3 = -400/27 < 0$. The cubic formula gives a messy answer: the roots are

$$g + h, \quad \zeta g + \zeta^2 h, \quad \zeta^2 g + \zeta h,$$

where $g^3 = \frac{1}{2}(-6 + \sqrt{-400/27})$ and $h^3 = \frac{1}{2}(-6 - \sqrt{-400/27})$. Something strange has happened. There are three curious equations saying that each of 1, 2, and $-3$ is equal to one of the messy expressions displayed above; thus,

$$\zeta \sqrt[3]{\frac{1}{2}(-6 + \sqrt{-400/27})} + \zeta^2 \sqrt[3]{\frac{1}{2}(-6 - \sqrt{-400/27})}$$

is equal to 1, 2, or $-3$! Aside from the complex cube roots of unity, this expression involves square roots of the negative number $-400/27$.

Until the Middle Ages, mathematicians had no difficulty in ignoring negative numbers or square roots of negative numbers when dealing with quadratic equations. For example, we have already mentioned that we ignored a negative root of a quadratic when using Galileo's equations to find the distance traveled by a shell fired from a hilltop. As a second example, consider the problem, mentioned at the beginning of this chapter, of finding the sides $x$ and $y$ of a rectangle having area $A$ and perimeter $p$. The equations

$$xy = A \quad \text{and} \quad 2x + 2y = p$$

lead to the quadratic equation $2x^2 - px + 2A = 0$, and the quadratic formula gives the roots

$$x = \frac{1}{4}\left(p \pm \sqrt{p^2 - 16A}\right).$$

If $p^2 - 16A \geq 0$, one has found $x$ (and $y$); if $p^2 - 16A < 0$, one merely says that there is no rectangle whose perimeter and area are in this relation. But the cubic formula does not allow one to discard "imaginary" roots, for we have just seen that an "honest" real and positive root, even a positive integer, can appear in terms of complex numbers. (We saw a similar phenomenon in Theorem 1.17, where the integer terms of the Fibonacci sequence were given in terms of the irrational number $\sqrt{5}$.) A nonconstant polynomial $f(x)$ with rational coefficients is called *irreducible* if it is has no factorization $f(x) = g(x)h(x)$ in which $g(x)$ and $h(x)$ are nonconstant polynomials having rational coefficients. One can prove the *casus irreducibilis*: If $f(x)$ is an irreducible cubic polynomial having all real roots, then there is no formula for a root of $f(x)$ in terms of real radicals; that is, for these polynomials, there is no way to manipulate the cubic formula to avoid using complex numbers.

The Pythagoreans in ancient Greece considered *number* to mean positive integer. By the Middle Ages, *number* came to mean positive real number (although there was little understanding then of what real numbers are). The importance of the cubic formula in the history of mathematics is that it forced mathematicians to take both complex numbers and negative numbers seriously.

The physicist R. P. Feynman (1918–1988), one of the first winners of the annual Putnam national mathematics competition (and also a Nobel laureate in physics), suggested another possible value of the cubic formula. As we mentioned at the beginning of this chapter, the cubic formula was found in 1515, a time of great change. One of the factors contributing to the Dark Ages was an almost slavish worship of the classical Greek and Roman civilizations. It was believed that that earlier era had been the high point of man's accomplishments; contemporary man was inferior to his forebears (a world view opposite to the modern one of continual progress!). The cubic formula was essentially the first instance of a mathematical formula unknown to the ancients, and so it may well have been a powerful example showing that sixteenth century man was the equal of his ancestors.

The quartic formula was discovered by Lodovici Ferrari (1522–1565) in the early 1540's; we present the derivation given by Descartes in 1637.

**Theorem 4.23 (Quartic Formula).** There is a method to compute the four roots of a quartic

$$f(X) = X^4 + bX^3 + cX^2 + dX + e.$$

*Proof.* The reader may generalize Lemma 4.21 as follows. The substitution $X = x - \frac{1}{4}b$ yields a *reduced* quartic

$$x^4 + qx^2 + rx + s;$$

that is, it has no $x^3$ term; moreover, if a number $u$ is a root of the reduced polynomial, then $u - \frac{1}{4}b$ is a root of the original.

Factor the reduced quartic into quadratics:

$$x^4 + qx^2 + rx + s = (x^2 + jx + \ell)(x^2 + kx + m).$$

If $j$, $k$, $\ell$, and $m$ can be found, then the quadratic formula can be used to find the roots of the reduced quartic.

The coefficient of $x^3$ on the right hand side is $j + k$, while the coefficient of $x^3$ in the reduced quartic on the left hand side is 0; hence, $k = -j$ and

$$x^4 + qx^2 + rx + s = (x^2 + jx + \ell)(x^2 - jx + m).$$

Expanding the right hand side and equating the coefficients of $x^2$, the coefficients of $x$, and the constant terms, one obtains

$$\left.\begin{aligned} m + \ell - j^2 &= q; \\ j(m - \ell) &= r; \\ \ell m &= s. \end{aligned}\right\} \tag{11}$$

Add and subtract the top two equations in Eqs. (11) to get

$$\left.\begin{aligned} 2m &= j^2 + q + r/j; \\ 2\ell &= j^2 + q - r/j. \end{aligned}\right\} \tag{12}$$

Now substitute these into the bottom equation of Eqs. (11):

$$\begin{aligned} 4s \;=\; 4\ell m &= \left( j^2 + q + \frac{r}{j} \right)\left( j^2 + q - \frac{r}{j} \right) \\ &= (j^2 + q)^2 - \frac{r^2}{j^2} \\ &= j^4 + 2j^2 q + q^2 - \frac{r^2}{j^2}. \end{aligned}$$

Clearing denominators and transposing gives

$$j^6 + 2qj^4 + (q^2 - 4s)j^2 - r^2 = 0, \tag{13}$$

a cubic equation in $j^2$. The cubic formula allows one to solve for $j^2$, and one then finds $\ell$ and $m$ using Eqs. (12).  •

The quartic formula appears in Cardano's book, but it is given much less attention there than the cubic formula. The reason given is that cubic polynomials have an interpretation as volumes, whereas quartic polynomials have no such obvious justification. We quote Cardano.

> As the first power refers to a line, the square to a surface, and the cube to a solid body, it would be very foolish for us to go beyond this point. Nature does not permit it. Thus, ... all those matters up to and including the cubic are fully demonstrated, but for the others which we will add, we do not go beyond barely setting out.

Another reason for not emphasizing the quartic is that it involves long calculations.

***Example.*** Consider

$$x^4 - 2x^2 + 8x - 3 = 0,$$

so that $q = -2, r = 8$, and $s = -3$. If we factor this reduced quartic into $(x^2 + jx + \ell)(x^2 - jx + m)$, then Eq. (13) gives

$$j^6 - 4j^4 + 16j^2 - 64 = 0.$$

One could use the cubic formula to find $j^2$, but this would be very tedious, for one must first get rid of the $j^4$ term before doing the rest of the calculations. It is simpler, in this case, to observe that $j = 2$ is a root, for the equation can be rewritten

$$j^6 - 4j^4 + 16j^2 - 64 = j^6 - 2^2 j^4 + 2^4 j^2 - 2^6 = 0$$

(many elementary texts are fond of saying, in such circumstances, that $j = 2$ is found "by inspection"). We now find $\ell$ and $m$ using Eqs. (12).

$$2\ell = 4 - 2 + \tfrac{8}{2} = 6$$

$$2m = 4 - 2 - \tfrac{8}{2} = -2.$$

Thus, the original quartic factors into

$$(x^2 + 2x + 3)(x^2 - 2x - 1).$$

The quadratic formula now gives the roots of the quartic:

$$-1+i\sqrt{2}, -1-i\sqrt{2}, 1+i\sqrt{2}, 1-i\sqrt{2}.$$

Do not be misled by this example; it is difficult to find a quartic whose roots are given by the quartic formula in recognizable form. For example, the quartic formula gives complicated versions of the roots of

$$x^4 - 25x^2 + 60x - 36 = (x-1)(x-2)(x-3)(x+6),$$

as the reader may check. See Exercise 4.41 for another bad example.

There is a derivation of the quadratic formula in the spirit of the derivations of the cubic and quartic formulas. Given

$$f(X) = X^2 + bX + c,$$

the substitution $X = x - \frac{1}{2}b$ yields the reduced quadratic

$$\begin{aligned}
\tilde{f}(x) &= (x - \tfrac{1}{2}b)^2 + b(x - \tfrac{1}{2}b) + c \\[2mm]
&= x^2 - bx + \tfrac{1}{4}b^2 + bx - \tfrac{1}{2}b^2 + c \\[2mm]
&= x^2 + c - \tfrac{1}{4}b^2.
\end{aligned}$$

Therefore, a root $u$ of $\tilde{f}(x)$ satisfies

$$u^2 = \tfrac{1}{4}b^2 - c = \tfrac{1}{4}(b^2 - 4c);$$

that is,

$$u = \pm \tfrac{1}{2}\sqrt{b^2 - 4c}.$$

It is easy to see, as in Lemma 4.21, that if $u$ is a root of $\tilde{f}(x)$, then $u - \frac{1}{2}b$ is a root of $f(x)$; that is, the roots of $f(x)$ are

$$u = -\tfrac{1}{2}b \pm \tfrac{1}{2}\sqrt{b^2 - 4c} = \tfrac{1}{2}\left[-b \pm \sqrt{b^2 - 4c}\right].$$

It is now very tempting, as it was for our ancestors, to seek the roots of a quintic $g(X) = X^5 + bX^4 + cX^3 + dX^2 + eX + f$ (of course, they hoped to find the roots of polynomials of any degree). Begin by making the substitution $X = x - \frac{1}{5}b$ to eliminate the quartic term. It is natural to expect that some further ingenious substitution together with the formulas for the roots of lower degree polynomials would yield the roots of $g(X)$, but quintics resisted all such attempts. After almost three hundred years, in 1799, P. Ruffini (1765–1822) published a proof that there is no formula analogous to the quadratic, cubic, and quartic formulas (i.e., formulas involving only the operations of addition, subtraction, multiplication, division, and extraction of roots) which gives the roots of a quintic polynomial in terms of its coefficients. While we can now see that Ruffini's proof was essentially correct, it does have gaps, and it was not accepted by his contemporaries. In 1824, N. H. Abel (1802–1829) gave a proof (essentially correct) which was accepted. Even though there is no formula (as described above) that gives the roots of the general quintic, it is possible to give a formula for the roots of some quintics, e.g., $x^5 - 1$. In 1831, E. Galois (1811–1832), only nineteen years old, was able to give a necessary and sufficient condition that a polynomial (of any degree) admit a formula for its roots that generalizes the quadratic, cubic, and quartic formulas. In so doing, he invented the Theory of Groups as well as what is nowadays called Galois Theory.

    A formula involving extraction of roots is not necessarily the simplest way to find the roots of a cubic. We shall now replace the operations of extraction of roots, which are, after all, "infinitary" in the sense that their evaluation requires limits, by evaluation of cosines in order to find the roots of a cubic $x^3 + qx + r$ in the case when all its roots are real. We remark that Newton's method, in calculus, shows how to find real roots as limits of sequences.

    There is a way to determine, in advance, whether or not a reduced cubic has all real roots. If the roots of a polynomial $f(x)$ are $\alpha_1, \alpha_2, \dots, \alpha_n$, that is, if $f(x) = a(x - \alpha_1)(x - \alpha_2) \cdots (x - \alpha_n)$ for some number $a$, define a number

$$\Delta = \prod_{1 \le i < j \le n} (\alpha_i - \alpha_j).$$

If $n = 2$, then $\Delta = \alpha_1 - \alpha_2$, and if $n = 3$, then $\Delta = (\alpha_1 - \alpha_2)(\alpha_1 - \alpha_3)(\alpha_2 - \alpha_3)$. Thus, $\Delta$ is the product of all the differences of pairs of roots. Now the number $\Delta$ depends not only on the roots themselves but also on the order in which they are listed. For example, if $n = 2$, let the two roots be $\alpha$ and $\beta$. There are two possible $\Delta$'s, namely, $\alpha - \beta$ and $\beta - \alpha = -(\alpha - \beta)$. If $n = 3$ and the roots are $\alpha, \beta, \gamma$, then one choice of $\Delta$ is $(\alpha - \beta)(\alpha - \gamma)(\beta - \gamma)$; another choice, arising from the ordering $\alpha, \gamma, \beta$, is $(\alpha - \gamma)(\alpha - \beta)(\gamma - \beta)$. In this case, the two $\Delta$'s are the same except for sign and, indeed, it is not difficult to see, more generally, that the $\Delta$'s arising from any two orderings of $\alpha_1, \alpha_2, \dots, \alpha_n$ either are equal or have opposite sign [for both numbers are products involving the same factors, perhaps with some factors $\alpha_i - \alpha_j$ in one product replaced by $\alpha_j - \alpha_i = -(\alpha_i - \alpha_j)$ in the other]. Squaring $\Delta$ eliminates this ambiguity.

**Definition.** Let $f(x)$ be a polynomial having $n$ roots $\alpha_1, \alpha_2, \ldots, \alpha_n$; that is, $f(x) = a(x - \alpha_1)(x - \alpha_2) \cdots (x - \alpha_n)$ for some number $a \neq 0$. The *discriminant* of $f(x)$ is the number

$$D = \Delta^2 = \left[ \prod_{1 \leq i < j \leq n} (\alpha_i - \alpha_j) \right]^2.$$

As an example, let $f(x) = x^2 + bx + c$. The quadratic formula gives the roots as $\alpha = \frac{1}{2}(-b + \sqrt{b^2 - 4c})$ and $\beta = \frac{1}{2}(-b - \sqrt{b^2 - 4c})$. Therefore, $\Delta = \alpha - \beta = \sqrt{b^2 - 4c}$, and so the discriminant is $b^2 - 4c$.

The discriminant of $f(x)$ is 0 if and only if $f(x)$ has repeated roots. But this is of no use to us unless we can compute the discriminant, as we have just done for quadratic polynomials, without having first to compute the roots themselves.

Let us now concentrate on cubic polynomials. If $f(x) = x^3 + bx^2 + cx + d$, then we claim that its reduced polynomial $\tilde{f}(x) = x^3 + qx + r$ has the same discriminant as $f(x)$. We saw, in Lemma 4.21, that if $\alpha, \beta, \gamma$ are the roots of $\tilde{f}(x)$, then $\alpha - \frac{1}{3}b$, $\beta - \frac{1}{3}b$, and $\gamma - \frac{1}{3}b$ are the roots of $f(x)$. Because $(\alpha - \frac{1}{3}b) - (\beta - \frac{1}{3}b) = \alpha - \beta$, etc., it follows that both $f(x)$ and $\tilde{f}(x)$ have the same $\Delta$, and, hence, they have the same discriminant $\Delta^2$. We are going to use the explicit formula in Theorem 4.22 [giving the roots of a cubic $f(x)$] to give a formula for the discriminant of $f(x)$.

**Lemma 4.24.** For any numbers $g$ and $h$,

$$(g - h)(g - \zeta h)(g - \zeta^2 h) = g^3 - h^3,$$

where $\zeta = e^{2\pi i/3}$ is a cube root of unity.

*Proof.* The formula holds if $h = 0$, and so we may assume that $h \neq 0$. The identity $x^3 - 1 = (x - 1)(x - \zeta)(x - \zeta^2)$ holds, in particular, for $x = g/h$. Hence,

$$\frac{g^3}{h^3} - 1 = \left( \frac{g}{h} - 1 \right) \left( \frac{g}{h} - \zeta \right) \left( \frac{g}{h} - \zeta^2 \right);$$

multiplying both sides by $h^3$ gives the result. •

***Lemma 4.25.*** The discriminant $D$ of $f(x) = x^3 + qx + r$ is

$$D = -27r^2 - 4q^3.$$

*Proof.* By the cubic formula, the roots of $f(x)$ are

$$\alpha = g + h, \beta = \zeta g + \zeta^2 h, \text{ and } \gamma = \zeta^2 g + \zeta h.$$

Because $\zeta^3 = 1$, we have

$$
\begin{aligned}
\alpha - \beta &= g + h - \zeta g - \zeta^2 h \\[6pt]
&= (g - \zeta^2 h) - (\zeta g - h) \\[6pt]
&= (g - \zeta^2 h) - \zeta(g - \zeta^2 h) \\[6pt]
&= (g - \zeta^2 h)(1 - \zeta).
\end{aligned}
$$

Similar calculations give

$$\alpha - \gamma = g + h - \zeta^2 g - \zeta h = (g - \zeta h)(1 - \zeta^2)$$

and

$$\beta - \gamma = \zeta g + \zeta^2 h - \zeta^2 g - \zeta h = (g - h)\zeta(1 - \zeta).$$

It follows that

$$\Delta = (g - h)(g - \zeta h)(g - \zeta^2 h)\zeta(1 - \zeta^2)(1 - \zeta)^2.$$

Now $\zeta(1 - \zeta^2)(1 - \zeta)^2 = 3i\sqrt{3}$, by Exercise 4.25, so that

$$D = \Delta^2 = -27[(g - h)(g - \zeta h)(g - \zeta^2 h)]^2.$$

By the lemma,

$$D = -27(g^3 - h^3)^2 = -27\left[\sqrt{R}\right]^2 = -27R,$$

where $R = r^2 + \frac{4}{27}q^3$. Hence, $D = -27r^2 - 4q^3$.   •

*Remark.* There is a formula for the discriminant of a quartic, but it is rather complicated.

For example, we see that $f(x) = x^3 - 3x + 2$ has repeated roots, for its discriminant is 0.

The quadratic formula shows that the roots of a quadratic are real if and only if its discriminant $\geq 0$. The next result shows that the same is true of a cubic.

**Theorem 4.26.** If $q$ and $r$ are real, then all the roots of $f(x) = x^3 + qx + r$ are real if and only if its discriminant $D \geq 0$; that is, $-27r^2 - 4q^3 \geq 0$.

*Proof.* If the roots $\alpha$, $\beta$, and $\gamma$ of $f(x)$ are all real, then

$$\Delta = (\alpha - \beta)(\alpha - \gamma)(\beta - \gamma)$$

is real, and hence the discriminant $D = \Delta^2$ is a nonnegative real number. Therefore, $-27r^2 - 4q^3 \geq 0$.

For the converse, we show that if one root $\beta$ is not real, then the discriminant $D < 0$. If $\beta = s + it$ with $t \neq 0$, then Exercise 4.15 says that the other two roots are $\gamma = s - it = \overline{\beta}$ and a real number $\alpha$. Thus,

$$
\begin{aligned}
\Delta &= (\alpha - \beta)(\alpha - \gamma)(\beta - \gamma) \\[2mm]
&= (\alpha - s - it)(\alpha - s + it)(s + it - (s - it)) \\[2mm]
&= (-2ti)[(\alpha - s)^2 + t^2].
\end{aligned}
$$

It follows that

$$
\begin{aligned}
\Delta^2 &= (-2ti)^2[(\alpha - s)^2 + t^2]^2 \\[2mm]
&= 4t^2 i^2 [(\alpha - s)^2 + t^2]^2 \\[2mm]
&= -4t^2[(\alpha - s)^2 + t^2]^2 < 0,
\end{aligned}
$$

for $\alpha$, $s$, and $t$ are real numbers and $t \neq 0$. Hence, if $f(x)$ has a nonreal root, then $D = \Delta^2 = -27r^2 - 4q^3 < 0$, as desired.  •

We are now going to find a trigonometric formula for the roots of a cubic. By Corollary 4.16(ii), we have

$$\cos 3\alpha = 4\cos^3 \alpha - 3\cos \alpha.$$

It follows that one root of the cubic

$$g(y) = y^3 - \tfrac{3}{4}y - \tfrac{1}{4}\cos 3\alpha \tag{14}$$

is $u = \cos\alpha$. By Exercise 4.35 below, the other two roots of this particular cubic are $u = \cos(\alpha + 120°)$ and $u = \cos(\alpha + 240°)$.

Now let $f(x) = x^3 + qx + r$ be a cubic all of whose roots are real (Theorem 4.26 gives a way of checking when this is the case). We want to force this equation to look like Eq. (14). If $v$ is a root of $f(x)$, set

$$v = tu,$$

where $t$ and $u$ will be chosen in a moment. Substituting,

$$0 = f(v) = t^3 u^3 + qtu + r,$$

and so

$$u^3 + \left(\frac{q}{t^2}\right)u + \frac{r}{t^3} = 0.$$

If we can choose $t$ so that $q/t^2 = -\tfrac{3}{4}$ and $r/t^3 = -\tfrac{1}{4}\cos 3\alpha$ for some $\alpha$, then $u$ is a root of $g(y)$; that is, $u = \cos\alpha$, $u = \cos(\alpha + 120°)$, or $u = \cos(\alpha + 240°)$, and so the roots of $f(x)$ are

$$v = tu = t\cos\alpha, \quad t\cos(\alpha + 120°), \quad \text{and} \quad t\cos(\alpha + 240°).$$

The equation $q/t^2 = -\tfrac{3}{4}$ gives $t^2 = -4q/3$, and so

$$t = \sqrt{-4q/3}. \tag{15}$$

Because all the roots are real, Theorem 4.26 gives $-27r^2 - 4q^3 \geq 0$, from which it follows that

$$-4q^3 \geq 27r^2;$$

as the right side is positive, $-4q^3$ is positive, and so $q$ must be negative. Therefore, $t = \sqrt{-4q/3}$ is a real number. The second equation gives

$$\cos 3\alpha = -\frac{4r}{t^3},$$

and this determines $\alpha$ provided that $|-4r/t^3| \leq 1$. Since $27r^2 \leq -4q^3$, we have $9r^2/q^2 \leq -4q/3$; taking square roots,

$$\left|\frac{3r}{q}\right| \leq \sqrt{\frac{-4q}{3}} = t,$$

by Eq. (15). Now $t^2 = -4q/3$, and so

$$\left|\frac{-4r}{t^3}\right| = \left|\frac{-4r}{(-4q/3)t}\right| = \left|\frac{3r}{qt}\right| \leq \frac{t}{t} = 1,$$

as desired. We have proved the following theorem of F. Viète.

**Theorem 4.27 (Viète).** Let $f(x) = x^3 + qx + r$ be a cubic polynomial all of whose roots are real. If $t = \sqrt{-4q/3}$ and $\cos 3\alpha = -4r/t^3$, then the roots of $f(x)$ are

$$t \cos\alpha, \quad t\cos(\alpha + 120°), \quad \text{and} \quad t\cos(\alpha + 240°).$$

**Example.** Consider the cubic $f(x) = x^3 - 7x + 6 = (x-1)(x-2)(x+3)$ that was discussed earlier; of course, its roots are $1, 2$, and $-3$ (had we not known the roots, we would have computed the discriminant: $q = -7, r = 6$, and $D = -27r^2 - 4q^3 = -27 \times 36 - 4 \times (-343) = 400 > 0$). The cubic formula gives rather complicated expressions for these roots in terms of cube roots of complex numbers involving $\sqrt{-400/27}$. Let us now find the roots of $f(x)$ using Theorem 4.27 (which applies because the roots of $f(x)$ are all real). We first compute $t$ and $\alpha$:

$$t = \sqrt{-4q/3} = \sqrt{-4(-7)/3} = \sqrt{28/3} \approx 3.05505$$

and

$$\cos 3\alpha = -4r/t^3 \approx -24/(3.05505)^3 \approx -.841698;$$

your favorite computer gives $3\alpha = \cos^{-1}(-.841698) \approx 147.3°$, and so

$$\alpha \approx 49.1°.$$

Hence, the roots of the cubic are, approximately,

$$3.05505 \cos 49.1° \approx 3.05505 \times .654741$$

$$\approx 2.000266 \approx 2,$$

$$3.05505 \cos 169.1° \approx 3.05505 \times -.981959$$

$$\approx -2.999934 \approx -3,$$

and

$$3.05505 \cos 289.1° \approx 3.05505 \times .327218$$

$$\approx 0.999667 \approx 1.$$

*Remark 1.* When solving the equation $\cos 3\alpha = -4r/t^3$ for $3\alpha$, one may assume that $3\alpha$ is in the first quadrant when $-4r/t^3$ is positive and in the second quadrant when $-4r/t^3$ is negative. See Exercise 4.31 for details.

*Remark 2.* When dealing with a cubic polynomial all of whose coefficients are integers, one should first determine whether or not it has any rational roots (see Theorem B on page 204).

*Remark 3.* If $f(x) = x^3 + qx + r$ has non-real roots, then a modification of Viète's method, using the hyperbolic functions $\cosh(x)$ and $\sinh(x)$, gives a formula for its real root. Recall that $\cosh(x) = \frac{1}{2}(e^x + e^{-x})$. There is a triple angle formula for cosh analogous to the triple angle formula for cosine:

$$\cosh(3\alpha) = 4\cosh^3(\alpha) - 3\cosh(\alpha).$$

This is easily proved: expand, then simplify $4[\frac{1}{2}(e^x + e^{-x})]^3 - 3[\frac{1}{2}(e^x + e^{-x})]$, and obtain $\frac{1}{2}(e^{3x} + e^{-3x})$. Therefore, $h(y) = y^3 - \frac{3}{4}y - \frac{1}{4}\cosh(3\alpha)$ has $u = \cosh(\alpha)$ as a root. To force $f(x) = x^3 + qx + r$ to look like $h(y)$, we write the real root $v$ of $f(x)$ as $v = tu$. As in the proof of Viète's theorem, we want $t^2 = -4q/3$ and $\cosh(3\alpha) = -4r/t^3$. If $-4q/3 \geq 0$, then $t$ is real. Using the discriminant condition $-27q^3 - 4r^2 < 0$, one can show (in a manner similar to that in the proof of Viète's theorem) that $-4r/t^3 \geq 1$. Now $\cosh(x) \geq 1$ for all $x$, and there is $\beta$ with $\cosh(\beta) = -4r/t^3$. It follows that the real root of $f(x)$ is given by

$$v = t\cosh(\beta/3),$$

where $t^2 = -4q/3 \geq 0$. [Of course, the other two (complex) roots of $f(x)$ are the roots of the quadratic $f(x)/(x - v)$.]

To treat the remaining possibility $-4q/3 < 0$, we use the hyperbolic sine:

$$\sinh(x) = \tfrac{1}{2}(e^x - e^{-x}).$$

There is a triple angle formula

$$\sinh(3\alpha) = 4\sinh^3(\alpha) + 3\sinh(\alpha),$$

so that $u = \sinh(\alpha)$ is a root of $k(y) = y^3 + \tfrac{3}{4}y - \tfrac{1}{4}\sinh(3\alpha)$. To force $f(x) = x^3 + qx + r$ to look like $k(y)$, we write the real root of $f(x)$ as $v = tu$, where $t^2 = 4q/3$ and $\sinh(3\alpha) = -4r/t^3$. We are now assuming that $-4q/3 < 0$, and so there is a real square root $t$ of $4q/3$. It is known that the second equation can always be solved: there is a number $\gamma$ with $\sinh(\gamma) = -4r/t^3$. Hence, the real root of $f(x)$ is

$$v = t\sinh(\gamma/3).$$

### Exercises

**4.31.** Assume that $0 \leq 3\alpha < 360°$.
(i) If $\cos 3\alpha$ is positive, show that there is an acute angle $\beta$ with $3\alpha = 3\beta$ or $3\alpha = 3(\beta + 90°)$, and that the sets of numbers

$$\cos\beta, \cos(\beta + 120°), \cos(\beta + 240°)$$

and

$$\cos(\beta + 90°), \cos(\beta + 210°), \cos(\beta + 330°)$$

coincide.
(ii) If $\cos 3\alpha$ is negative, there is an acute angle $\beta$ with $3\alpha = 3(\beta + 30°)$ or $3\alpha = 3(\beta + 60°)$, and that the sets of numbers

$$\cos(\beta + 30°), \cos(\beta + 150°), \cos(\beta + 270°)$$

and

$$\cos(\beta + 60°), \cos(\beta + 180°), \cos(\beta + 270°)$$

coincide.

**4.32.** Consider the polynomial $f(X) = X^3 + X^2 - 36$ that arose in the castle problem in Chapter 2.

(i) Show that 3 is a root of $f(X)$ and find the other two roots as roots of the quadratic $f(X)/(X-3)$.                Answer: $-2 \pm i\sqrt{8}$.

(ii) Use the cubic formula to find the root 3 of $f(X)$. (This is a very tedious exercise; it gives another reason why the cubic formula is not popular.)

(iii) Show that the discriminant of $f(X)$ is negative, and find its real root (which is $\frac{10}{3}$) using cosh. (Hint: If you do not have a calculator or table of hyperbolic cosines, you can use the definition of cosh; in particular, $\cosh(x) \approx \frac{1}{2}e^x$ for "large" $x$.)

**4.33.** (i) Show, for all $a$ and $b$ and for all $j \geq 1$, that

$$a^j - b^j = (a-b)(a^{j-1} + a^{j-2}b + a^{j-3}b^2 + \cdots + ab^{j-2} + b^{j-1}).$$

(ii) Let $f(x) = c_n x^n + c_{n-1} x^{n-1} + \cdots + c_1 x + c_0$ be a polynomial of degree $n$. For any number $u$, show there is some polynomial $q(x)$ of degree $n-1$ with

$$f(x) = (x - u)q(x) + f(u).$$

[Hint:

$$f(x) - f(u) = \sum_{j=0}^{n} c_j x^j - \sum_{j=0}^{n} c_j u^j = \sum_{j=0}^{n} c_j(x^j - u^j);$$

now use part (i).]

(iii) Show that $u$ is a root of $f(x)$ if and only if $x - u$ is a factor of $f(x)$.

**4.34.** Show that every cubic polynomial $f(x) = x^3 + bx^2 + cx + d$ having real coefficients $b$, $c$, and $d$ has at least one real root. (Hint: Use Exercise 4.16.)

**4.35** Show that if $\cos 3\alpha = r$, then the roots of $4x^3 - 3x - r$ are

$$\cos(\alpha), \cos(\alpha + 120°), \cos(\alpha + 240°).$$

[Hint: $\cos 3\alpha = \cos(3[\alpha + 120°]) = \cos(3[\alpha + 240°])$.]

**4.36.** Find the roots of $f(x) = x^3 - 3x + 1$.

**4.37.** Find the roots of $f(x) = x^3 - 9x + 28$.                Answer: $-4, 2 \pm i\sqrt{3}$.

**4.38.** Find the roots of $f(x) = x^3 - 24x^2 - 24x - 25$.    Answer: $17, -\frac{1}{2} \pm i\frac{\sqrt{3}}{2}$.

**4.39.** (i) Find the roots of $f(x) = x^3 - 15x - 4$ using the cubic formula.
                Answer: $g = \sqrt[3]{2 + \sqrt{-121}}$ and $h = \sqrt[3]{2 - \sqrt{-121}}$.
(ii) Find the roots using the trigonometric formula.    Answer: $4, -2 \pm \sqrt{3}$.

**4.40.** Find the roots of $f(x) = x^3 - 6x + 4$.                Answer: $2, -1 \pm \sqrt{3}$.

**4.41.** Find the roots of $x^4 - 15x^2 - 20x - 6$.            Answer: $-3, -1, 2 \pm \sqrt{6}$.

## IRRATIONALITIES

It can be of no practical use to know that Pi is irrational, but if we can know, it surely would be intolerable not to know.

*E. C. Titchmarsh*

In this section, we are going to show that certain numbers are irrational. We begin with $e$, which we define as the sum of the series

$$e = 1 + 1 + \frac{1}{2!} + \cdots + \frac{1}{n!} + \cdots.$$

The following result appears in almost every calculus text (e.g., see Edwards & Penney, *Calculus*, page 622).

**Theorem A.** Let $a_1 > a_2 > a_3 > \cdots$ be a decreasing sequence of positive real numbers converging to 0. Then the alternating series $\sum_{i=0}^{\infty}(-1)^i a_i$ converges to a number $S$ with

$$a_1 - a_2 < S < a_1.$$

The following result was first proved by Euler in 1744; the proof we give is due to L. L. Pennisi in 1953.

**Theorem 4.28.** $e$ is irrational.

*Proof.* Suppose, on the contrary, that $e = k/m$, where $k$ and $m$ are positive integers. Then

$$e^{-1} = \frac{m}{k} = \sum_{n=0}^{\infty} \frac{(-1)^n}{n!},$$

and so

$$\frac{m}{k} - \sum_{n=0}^{k} \frac{(-1)^n}{n!} = \sum_{n>k} \frac{(-1)^n}{n!}.$$

Multiply both sides by $\pm k!$ (the sign chosen so that the series on the right side begins with a positive term) to obtain

$$\pm (k-1)!m - \sum_{n=0}^{k} \frac{(-1)^n k!}{n!} =$$

$$\frac{1}{k+1} - \frac{1}{(k+1)(k+2)} + \frac{1}{(k+1)(k+2)(k+3)} - \cdots. \quad (16)$$

Now the left side of Eq. (16) is an integer $S$, say, for $k!$ is a multiple of all denominators occurring in it. On the other hand, the alternating series on the right side of Eq. (16) converges to $S$, and Theorem A says that

$$\frac{1}{k+1} - \frac{1}{(k+1)(k+2)} < S < \frac{1}{k+1}.$$

The left side simplifies to $1/(k+2)$, so that

$$\frac{1}{k+2} < S < \frac{1}{k+1}.$$

But there are no integers between these two fractions, and we have reached a contradiction.  •

*Remark.* In 1873, C. Hermite (1822–1901) proved that $e$ is ***transcendental***; that is, there is no polynomial $f(x)$ with rational coefficients having $e$ as a root [were $e$ rational, then it would be a root of a polynomial with rational coefficients, namely, $f(x) = x - e$].

The proof of the next main theorem is adapted from the book by I. Niven and H. S. Zuckerman, *An Introduction to the Theory of Numbers*. When first learning trigonometry, the first few values of $\cos\theta$ one learns are $\cos 0 = 1$, $\cos(\frac{\pi}{6}) = \frac{\sqrt{3}}{2}$, $\cos(\frac{\pi}{4}) = \frac{\sqrt{2}}{2}$, $\cos(\frac{\pi}{3}) = \frac{1}{2}$, and $\cos(\frac{\pi}{2}) = 0$; three of these five are rational. We shall see how misleading this first impression is.

The following result will be needed.

***Theorem B.*** If $f(x) = x^n + a_{n-1}x^{n-1} + \cdots a_1 x + a_0$ is a polynomial with integer coefficients $a_{n-1}, \ldots, a_1, a_0$, then every rational root of $f(x)$ is an integer that is a divisor of $a_0$.

We leave the proof of Theorem B to the appropriate course (e.g., see [Rotman, p. 182]), but here is a nice consequence of it. If $a$ is an integer that is not an $n$th power, i.e., there is no integer $b$ with $a = b^n$, then $\sqrt[n]{a}$ is irrational. After all, $\sqrt[n]{a}$ is a root of $f(x) = x^n - a$, and Theorem B applies to say that $f(x)$ has no rational roots. This is far stronger than our earlier discussion when we proved, in Theorem 2.5, that $\sqrt{2}$ is irrational and, more generally, when we described, in Exercise 2.8, why $\sqrt{m}$ is irrational whenever $m$ is not a perfect square.

Theorem B would also have been useful when we were looking at the cubic $x^3 + x^2 - 36$ that arose in the castle problem in Chapter 2 and again in Exercise 4.32. Any rational roots of it are necessarily integers which are divisors of $-36$, and testing each of them in turn would have quickly given $x = 3$ as a root. Indeed, given any cubic or quartic with integer coefficients, one should first seek rational roots using Theorem B before embarking on the long journey using the cubic or quartic formulas.

The method of Diophantus described all those numbers $\theta$ for which both $\cos\theta$ and $\sin\theta$ are rational; we are now going to describe many $\theta$ for which both $\cos\theta$ and $\sin\theta$ are irrational.

De Moivre's theorem tells us, for every real number $\theta$, that $\cos n\theta$ is the real part of $(\cos\theta + i\sin\theta)^n$. The binomial theorem (Exercise 4.20) gives

$$(\cos\theta + i\sin\theta)^n = \sum_{r=0}^{n} \binom{n}{r} i^r \cos^{n-r}\theta \sin^r\theta.$$

Because only even powers of $i$ are real, it follows that $\cos n\theta$ is the sum of all those terms $\binom{n}{r} i^r \cos^{n-r}\theta \sin^r\theta$ for which $r$ is even; since $i^{2k} = (-1)^k$, it follows that $\cos n\theta$ is the sum of all terms of the form

$$\binom{n}{2k}(-1)^k \cos^{n-2k}\theta \sin^{2k}\theta.$$

But $\sin^2\theta = 1 - \cos^2\theta$, hence, $\sin^{2k}\theta = (1 - \cos^2\theta)^k$, and so $\cos n\theta$ is some polynomial $g_n(x)$, with integer coefficients, in powers of $x = \cos\theta$; that is, $\cos n\theta = g_n(\cos\theta)$. For example, Corollary 4.16 gives the formulas $\cos 2\theta = 2\cos^2\theta - 1$ and $\cos 3\theta = 4\cos^3\theta - 3\cos\theta$; thus, $g_2(x) = 2x^2 - 1$ and $g_3(x) = 4x^3 - 3x$. It is a good guess that the coefficient of $x^n$ in $g_n(x)$ is $2^{n-1}$, but in order to use Theorem B, we prefer a polynomial in which the coefficient of the highest power of $x$ is 1.

*Lemma 4.29.* For every $n \geq 1$, there is a polynomial with integer coefficients,

$$f_n(x) = x^n + a_{n-1}x^{n-1} + \cdots + a_1 x + a_0,$$

for which $2\cos n\theta = f_n(2\cos\theta)$.

*Proof.* The proof is by the second form of induction; the base step $n = 1$ is easy: define $f_1(x) = x$. Indeed, we may define $f_2(x) = x^2 - 2$, since $2\cos 2\theta = (2\cos\theta)^2 - 2 = 4\cos^2\theta - 2$.

Exercise 4.27 (= Exercise 2.51) gives the identity, for all $n \geq 1$,

$$2\cos(n + 1)\theta = (2\cos\theta)(2\cos n\theta) - 2\cos(n - 1)\theta.$$

For the inductive step, we may assume that $f_n(x)$ and $f_{n-1}(x)$ are known, and we define

$$f_{n+1}(x) = xf_n(x) - f_{n-1}(x).$$

Setting $x = 2\cos\theta$, we have

$$f_{n+1}(2\cos\theta) = (2\cos\theta)f_n(2\cos\theta) - f_{n-1}(2\cos\theta)$$

$$= (2\cos\theta)(2\cos n\theta) - 2\cos(n-1)\theta,$$

by the inductive hypothesis. Therefore, $f_{n+1}(2\cos\theta) = 2\cos(n+1)\theta$, by the cited exercise. $\bullet$

**Theorem 4.30.** If $\theta = r\pi$, where $r$ is a rational number, then the only rational values of $\cos(r\pi)$ and $\sin(r\pi)$ are $0$, $\pm\frac{1}{2}$, and $\pm 1$.

*Proof.* Because $r$ is a rational number, there is an integer $n$ so that $nr$ is an integer. With $f_n(x)$ as in the lemma and $x = r\pi$, we have

$$f_n(2\cos(r\pi)) = 2\cos(nr\pi) = \pm 2$$

(because $\cos(m\pi) = \pm 1$ for every integer $m$); it follows that $2\cos(r\pi)$ is a root of $f_n(x) \pm 2$. If $2\cos(r\pi)$ is rational, then Theorem B says it is an integer. But $-2 \le 2\cos(r\pi) \le 2$ [because $-1 \le \cos\theta \le 1$ always], and the only integers in this range are $\pm 1$, $0$, and $\pm 2$. It follows that the only possible rational values of $\cos(r\pi)$ are $\pm\frac{1}{2}$, $0$, and $\pm 1$.

Since $\sin(\theta) = \cos(\frac{1}{2}\pi - \theta)$, we have

$$\sin r\pi = \cos(\tfrac{1}{2}\pi - r\pi) = \cos([\tfrac{1}{2} - r]\pi).$$

It follows that $\sin(r\pi)$ takes on the same rational values as does cosine. $\bullet$

For example, both $\sin(36°)$ and $\cos(36°)$ are irrational.

Of course, the Intermediate Value Theorem says that $\cos\theta$ and $\sin\theta$ take on every value in the interval $[-1, 1]$.

We are now going to prove the irrationality of $\pi$, first proved by J. H. Lambert in 1767 (actually, we will prove that $\pi^2$ is irrational, which is a stronger result, first proved by A.-M. Legendre in 1794: were $\pi$ rational, then $\pi^2$ would also be rational). Before we do so, we must mention a much deeper theorem: in 1882, F. Lindemann (1852–1939) proved that $\pi$ is transcendental (this result implies that the classical Greek problem of "squaring the circle" is impossible to solve with only straightedge and compass). There is a proof of Lindemann's theorem in [Hadlock] that is accessible to beginning graduate students.

The coming proof, due to T. Estermann in 1965, is ingenious. The first lemma is reminiscent of the ratio test for convergence of series.

**Lemma 4.31.** If $\{u_n\}$ is a sequence of positive numbers with $u_{n+1}/u_n \to 0$, then $u_n \to 0$.

*Proof.* Because $u_{n+1}/u_n \to 0$, there is an integer $\ell$ with

$$\frac{u_{n+1}}{u_n} < \tfrac{1}{2}$$

for all $n \geq \ell$. We prove first, by induction on $m \geq 1$, that

$$u_{\ell+m} < (\tfrac{1}{2})^m u_\ell.$$

The base step is true, for $u_{\ell+1}/u_\ell < \tfrac{1}{2}$ gives $u_{\ell+1} < \tfrac{1}{2}u_\ell$ (because $u_\ell$ is positive). The inductive step is true, for if $u_{\ell+m} < (\tfrac{1}{2})^m u_\ell$, then $u_{\ell+m+1} < \tfrac{1}{2}u_{\ell+m}$, by the displayed inequality, and $\tfrac{1}{2}u_{\ell+m} < \tfrac{1}{2}[(\tfrac{1}{2})^m u_\ell] = (\tfrac{1}{2})^{m+1}u_\ell$.

Let $\varepsilon > 0$ be given. By Corollary 3.20, $(\tfrac{1}{2})^m u_\ell \to 0$ (where $\ell$ is fixed), so there is some integer $\ell'$ with $(\tfrac{1}{2})^m u_\ell < \varepsilon$ for all $m \geq \ell'$. It follows that $u_{\ell+m} < (\tfrac{1}{2})^m u_\ell < \varepsilon$ for all $m \geq \ell'$; that is, $u_k < \varepsilon$ for all $k = m+\ell \geq \ell'+\ell$. Therefore, $u_n \to 0$. •

If $\pi^2$ were rational, then there would be positive integers $a$ and $b$ with

$$\pi^2/4 = a/b$$

(we have normalized a bit for future convenience). Define functions

$$f_n(t) = \frac{(a - bt^2)^n}{n!},$$

for all $n = 0, 1, 2, \ldots$ .

**Lemma 4.32.** For all $n \geq 0$, the functions $f_n(t)$, satisfy the following properties.
(i) $f_n(0) = a^n/n!$ and $f_n(\tfrac{\pi}{2}) = 0$.
(ii) $f_n(t) > 0$ for all $t$ in $[0, \tfrac{\pi}{2})$.
(iii) $f_0(t) = 1$ for all $t$ and, if $n \geq 1$, then $f_n'(t) = -2bt f_{n-1}(t)$.
(iv) $f_n(t)$ is decreasing on $[0, \tfrac{\pi}{2}]$.

*Proof.* (i) Substitute into the definition of $f_n(t)$.

(ii) The sign of $f_n(t)$ is determined by the sign of $a - bt^2$. Now if $0 \le t < \frac{\pi}{2}$, then $0 \le t^2 < (\frac{\pi}{2})^2 = a/b$ (this is the reason for the normalization), hence $0 \le bt^2 < a$, and so $a - bt^2 > a - a = 0$.

(iii) If $n = 0$, then $f_0(t) = (a - bt^2)^0/0! = 1$. If $n \ge 1$, then

$$f_n'(t) = n(a - bt^2)^{n-1}(-2bt)/n! = -2bt f_{n-1}(t).$$

(iv) Because $f_{n-1}(t) \ge 0$ on the interval $[0, \frac{\pi}{2}]$, by (ii), it follows that $f_n'(t)$ is negative, and so $f_n(t)$ is decreasing.   •

**Definition.** Define numbers $I_n$, for all $n \ge 0$, by

$$I_n = \int_0^{\pi/2} f_n(t) \cos t \, dt.$$

**Lemma 4.33.** The numbers $I_0, I_1, I_2, \ldots, I_n, \ldots$ satisfy the following properties.

(i) $I_0 = 1$ and $I_1 = 2b$.
(ii) $I_n > 0$ for all $n \ge 0$.
(iii) $I_n \le \frac{\pi}{2} a^n/n!$ for all $n \ge 0$.

*Proof.* (i) $I_0 = 1$ because $f_0(t) = 1$ for all $t$ and $\int_0^{\pi/2} \cos t \, dt = 1$. The value for $I_1 = \int_0^{\pi/2} f_1(t) \cos t \, dt = \int_0^{\pi/2} (a - bt^2) \cos t \, dt$ is found by integrating by parts[17] twice. Setting $u = f_1(t)$ and $dv = \cos t dt$, we have, using Lemma 4.32(iii),

$$\int_0^{\pi/2} (a - bt^2) \cos t \, dt \;=\; f_1(t) \sin t \Big|_0^{\pi/2} - \int_0^{\pi/2} f_1'(t) \sin t \, dt$$

$$=\; f_1(t) \sin t \Big|_0^{\pi/2} + \int_0^{\pi/2} 2bt \sin t \, dt.$$

Now $f_1(t) \sin t \big|_0^{\pi/2} = f_1(\frac{\pi}{2}) \sin(\frac{\pi}{2}) - f_1(0) \sin 0 = 0$, and a second integration by parts gives

$$\int_0^{\pi/2} 2bt \sin t \, dt = 2b \left( -t \cos t \Big|_0^{\pi/2} + \int_0^{\pi/2} \cos t \, dt \right) = 2b.$$

---

[17]Recall that integration by parts is the formula $\int_a^b u \, dv = uv \big|_a^b - \int_a^b v \, du$.

(ii) By Lemma 4.32(ii), $f_n(t) > 0$ on the interval $[0, \frac{\pi}{2})$; since $\cos t > 0$ on this same half open interval, it follows that

$$I_n = \int_0^{\pi/2} f_n(t) \cos t \, dt > 0$$

(the area under the curve $y = f_n(t) \cos t$ must be positive).

(iii) If $n = 0$, then $I_n = 1 \leq \frac{\pi}{2}a^0/0! = \frac{\pi}{2}$. If $n \geq 1$, then $f_n(t)$ is decreasing on $[0, \frac{\pi}{2}]$, by Lemma 4.32(iv), and so its largest value $M$ on this interval is $f_n(0) = a^n/n!$. Since the length of the interval is $\frac{\pi}{2}$, it follows that

$$I_n \leq \frac{\pi}{2}M = \frac{\pi}{2}\frac{a^n}{n!}. \quad \bullet$$

The last (computational) lemma shows that the numbers $I_n$ are integers that satisfy a recurrence relation.

**Lemma 4.34.** (i) If $g_n(t) = f_n(t) \sin t + f'_n(t) \cos t$, where $n \geq 2$, then

$$\left[ f_n(t) - (4n - 2)bf_{n-1}(t) + 4abf_{n-2}(t) \right] \cos t = g'_n(t).$$

(ii) For all $n \geq 2$,

$$I_n - (4n - 2)bI_{n-1} + 4abI_{n-2} = 0.$$

(iii) $I_n$ is a positive integer for all $n \geq 0$.

*Proof.* (i) From Lemma 4.32(iii), $f'_n(t) = -2btf_{n-1}(t)$ for $n \geq 1$, so that, for all $n \geq 2$, we have

$$f''_n(t) = -2bf_{n-1}(t) + 4b^2t^2 f_{n-2}(t).$$

Since $g'_n(t) = \left[ f_n(t) \sin t + f'_n(t) \cos t \right]'$, we have

$$
\begin{aligned}
g'_n(t) &= f_n(t) \cos t + f'_n(t) \sin t - f'_n(t) \sin t + f''_n(t) \cos t \\[2mm]
&= f_n(t) \cos t + f''_n(t) \cos t \\[2mm]
&= f_n(t) \cos t + [-2bf_{n-1}(t) + 4b^2t^2 f_{n-2}(t)] \cos t \\[2mm]
&= \left[ f_n(t) - 2bf_{n-1}(t) + 4b^2t^2 f_{n-2}(t) \right] \cos t.
\end{aligned}
$$

Finally, since $f_n(t) = (a - bt^2)^n/n!$, we have

$$(a - bt^2) f_{n-2}(t) = (n - 1) f_{n-2}(t);$$

transposing gives

$$bt^2 f_{n-2}(t) = af_{n-2}(t) - (n - 1) f_{n-1}(t).$$

Substituting into the equation for $g_n'(t)$ above gives the result:

$$
\begin{aligned}
g_n'(t) &= \left[ f_n(t) - 2bf_{n-1}(t) + 4b^2 t^2 f_{n-2}(t) \right] \cos t \\[2mm]
&= \left[ f_n(t) - 2bf_{n-1}(t) + 4b[af_{n-2}(t) - (n-1)f_{n-1}(t)] \right] \cos t \\[2mm]
&= \left[ f_n(t) - (4n - 2)bf_{n-1}(t) + 4abf_{n-2}(t) \right] \cos t.
\end{aligned}
$$

(ii) Integrate both sides of the equation above from 0 to $\frac{\pi}{2}$. The left side becomes $I_n - (4n - 2)bI_{n-1} + 4abI_{n-2}$, while the right side, by the fundamental theorem of calculus, is

$$g_n(t) \Big|_0^{\pi/2} = g_n\left(\tfrac{\pi}{2}\right) - g_n(0)$$

$$= f_n\left(\tfrac{\pi}{2}\right) \sin\left(\tfrac{\pi}{2}\right) + f_n'\left(\tfrac{\pi}{2}\right) \cos\left(\tfrac{\pi}{2}\right) - \left[ f_n(0) \sin 0 + f_n'(0) \cos 0 \right].$$

Now $f_n\left(\frac{\pi}{2}\right) = 0 = \cos\left(\frac{\pi}{2}\right) = \sin 0$, whereas Lemma 4.32(iii) says that $f_n'(t) = -2bt f_{n-1}(t)$, whence $f_n'(0) = 0$. Therefore, $g_n(t)\big|_0^{\pi/2} = 0$, as desired.

(iii). Now $I_n$ is positive, by Lemma 4.33(ii). We prove that $I_n$ is an integer by induction on $n \geq 0$. By Lemma 4.33(i), $I_0 = 1$ and $I_1 = 2b$. If $n \geq 2$, then the inductive hypothesis (second form of induction) gives $I_{n-1}$ and $I_{n-2}$ integers, so that $I_n = (4n - 2)bI_{n-1} - 4abI_{n-2}$ is also an integer. ●

**Theorem 4.35.** $\pi$ is irrational (indeed, $\pi^2$ is irrational).

**Proof.** If we set $u_n = \frac{\pi}{2}a^n/n!$, then

$$\frac{u_{n+1}}{u_n} = \frac{\frac{\pi}{2}a^{n+1}/(n+1)!}{\frac{\pi}{2}a^n/n!} = \frac{a}{n+1} \to 0.$$

By Lemma 4.31, $\frac{\pi}{2}a^n/n! \to 0$. But $I_n \le \frac{\pi}{2}a^n/n!$, by Lemma 4.33(iii), so that $I_n \to 0$, by Exercise 3.22 (the sandwich theorem), because $0 < I_n \le \frac{\pi}{2}a^n/n!$ for all $n$. Therefore, if $\varepsilon = \frac{1}{2}$, there is some $\ell$ with $I_n < \frac{1}{2}$ for all $n \ge \ell$, contradicting Lemma 4.34(iii) which says that $I_n$ is a positive integer for all $n \ge 0$. This contradiction shows that there can be no integers $a$ and $b$ with $\pi^2/4 = a/b$; that is, $\pi^2$ is irrational. •

## *Exercises*

**4.42.** (i) Show that

$$\cos 2\theta = \frac{1 - \tan^2 \theta}{1 + \tan^2 \theta}.$$

(ii) If $r$ is a rational number, show that the only rational values of $\tan(r\pi)$ are 0 and $\pm 1$.

**4.43.** (i) Give an example of two positive irrationals whose sum is rational.
(ii) Show that $\sqrt{2} + \sqrt{3}$ is a root of $x^4 - 10x^2 + 1$.
(iii) Use Theorem B to show that $\sqrt{2} + \sqrt{3}$ is irrational.

**4.44.** (i) Prove that $\log_5 6$ is irrational.
(ii) Prove that $\log_6 15$ is irrational.

## EPILOGUE

Our journey into mathematics is over. We began with the most elementary considerations of area and numbers and, by asking natural questions, we were led to quite interesting results. There are other journeys, either continuing further in the direction we have been going, investigating more deeply some of the subjects we have touched upon, or going out in different directions altogether. But no matter what direction you pursue, you will find that solutions to old questions generally beget new questions of their own. Journeys almost always write their own itinerary, and looking back from the journey's end, you find that you understand the world a little better than you did when setting out. Mathematics is a never-ending story.

## BIBLIOGRAPHY

Bartle, R. G., and Sherbert, D. R., *Introduction to Real Analysis,*
Second Edition, Wiley, 1992.

Beckmann, P., *A History of* $\pi$, Golem Press, 1971,
Barnes & Noble, 1993

Cajori, F., *A History of Mathematical Notation*, Open Court,
1928, Dover reprint, 1993.

Edwards, C.H.,Jr., and Penney, D.E., *Calculus with Analytic Geometry*,
Fourth Edition, Prentice-Hall, 1994.

Eves, H. W., *An Introduction to the History of Mathematics,*
Sixth Edition, Saunders, 1990.

Hadlock, C. R., *Field Theory and Its Classical Problems*, Carus
Math. Monographs 19, Math. Assoc. America, 1978.

Heath, T. L., *The Thirteen Books of Euclid's Elements*, Second
Edition, Cambridge University Press, 1926, Dover reprint, 1956.

Niven, I., and Zuckerman, H. S., *An Introduction to the Theory
of Numbers*, Third Edition, Wiley, 1972.

Pólya, G., and Latta, G., *Complex Variables*, Wiley, 1974.

Rotman, J. J., *A First Course in Abstract Algebra*, Prentice-Hall, 1996.

Silverman, J. H., and Tate, J., *Rational Points on Elliptic Curves,*
Springer-Verlag, 1992.

Stillwell, J., *Mathematics and Its History*, Springer-Verlag, 1989.

Tignol, J.-P., *Galois' Theory of Algebraic Equations*, Wiley, 1988.

van der Waerden, B. L., *Science Awakening*, Wiley, 1963.

# Glossary of Logic

Logic, like whiskey, loses its beneficial effect when taken in too large quantities.

*Lord Dunsany*

What follows is designed as a chatty dictionary, but instead of arranging the terms in alphabetical order, we set them down in a natural sequence. When looking for a particular term, it is best to find it in the Index (which is arranged alphabetically) before looking for it here.

### PROPOSITIONAL CALCULUS

A language involves sentences, connectives (e.g., *not, and, or*) and rules for forming sentences. But a language is more than these; some sentences must be true or false. In English, "I am a thousand years old" and "Every even integer $\geq 4$ is the sum of two primes" are sentences that are true or false (even if we can't decide which), but "My cat has six kittens" is neither true nor false, for I do not own a cat.

We are now going to abstract the notion of language. In contrast to spoken languages, we will describe a language, called the *propositional calculus*, in which every sentence is either true or false (once truth values are assigned to its component parts). A passage from this abstract language to an honest language is called an *interpretation.* In giving definitions within the propositional calculus, we will always be guided by what we want to hold in familiar interpretations.

We begin with *sentence symbols* $P$, $Q$, ..., each of which may be assigned a *truth value* $T$ (true) or $F$ (false). (There are languages, the so-called *many-valued logics*, in which there may be more than two truth values—one interpretation of such languages is that sentences have a certain probability of being true.) Two intertwined aspects in the construction of the propositional calculus are: Which strings of sentence symbols, parentheses, and connectives are formulas (sentences are usually called *formulas* in this context); What are the truth values of the formulas. We now introduce connectives that

217

are used to build formulas. Each sentence symbol is, by definition, a formula, and the idea now is to gradually build longer formulas from shorter ones by using connectives. We shall not give the details of this construction, for they are rather fussy, but here is the description of the connectives. Let $\varphi$ (lower case Greek *phi*) and $\psi$ (lower case Greek *psi*) be formulas.

*Negation*: The negation of a formula $\varphi$, pronounced "not $\varphi$," is a formula denoted by $\sim \varphi$.

We describe $\sim \varphi$ by saying when it is true and when it is false. If $\varphi$ is false, then $\sim \varphi$ is true, and if $\varphi$ is true, then $\sim \varphi$ is false. It is convenient to display this in a ***truth table***, where $T$ denotes "true" and $F$ denotes "false."

| $\varphi$ | $T$ | $F$ |
|---|---|---|
| $\sim \varphi$ | $F$ | $T$ |

*And*: The formula "$\varphi$ and $\psi$" is denoted by $\varphi \wedge \psi$; here is its truth table.

| $\varphi$ | $T$ | $T$ | $F$ | $F$ |
|---|---|---|---|---|
| $\psi$ | $T$ | $F$ | $T$ | $F$ |
| $\varphi \wedge \psi$ | $T$ | $F$ | $F$ | $F$ |

Thus, $\varphi \wedge \psi$ is true precisely when both $\varphi$ and $\psi$ are true.

*Or*: The formula "$\varphi$ or $\psi$" is denoted by $\varphi \vee \psi$; here is its truth table.

| $\varphi$ | $T$ | $T$ | $F$ | $F$ |
|---|---|---|---|---|
| $\psi$ | $T$ | $F$ | $T$ | $F$ |
| $\varphi \vee \psi$ | $T$ | $T$ | $T$ | $F$ |

Thus, $\varphi \vee \psi$ is true when at least one of $\varphi$ and $\psi$ is true. In particular, we have declared $\varphi \vee \psi$ to be true if both $\varphi$ and $\psi$ happen to be true. (There is another usage of the connective *or* in common parlance, to the effect that either $\varphi$ or $\psi$, but not both, is true (e.g., Take it or leave it!); this is the so-called ***exclusive or***, but we always mean *or* as defined by the truth table above.)

***If and only if***: The formula "$\varphi$ if and only if $\psi$" is denoted by $\varphi \Leftrightarrow \psi$; here is its truth table.

| $\varphi$ | $T$ | $T$ | $F$ | $F$ |
|---|---|---|---|---|
| $\psi$ | $T$ | $F$ | $T$ | $F$ |
| $\varphi \Leftrightarrow \psi$ | $T$ | $F$ | $F$ | $T$ |

Thus, $\varphi \Leftrightarrow \psi$ is true precisely when the truth values of $\varphi$ and $\psi$ agree. We say that two formulas $\varphi$ and $\psi$ are ***logically equivalent*** if they have the same truth tables. It is easy to see that $\varphi$ and $\psi$ are logically equivalent precisely when the formula $\varphi \Leftrightarrow \psi$ is a ***tautology*** (from the Greek word meaning *same statement*); that is, the formula is true for every assignment of truth values to the sentence symbols in it. Problems asking whether two particular formulas are logically equivalent are routine; just compare their truth tables. Here is an example. The formulas

$$\sim (\varphi \vee \psi) \quad \text{and} \quad [(\sim \varphi) \wedge (\sim \psi)]$$

are logically equivalent because their truth tables are the same:

| $\varphi$ | $T$ | $T$ | $F$ | $F$ |
|---|---|---|---|---|
| $\psi$ | $T$ | $F$ | $T$ | $F$ |
| $\sim (\varphi \vee \psi)$ | $F$ | $F$ | $F$ | $T$ |
| $(\sim \varphi) \wedge (\sim \psi)$ | $F$ | $F$ | $F$ | $T$ |

The formulas

$$\sim (\varphi \vee \psi) \Leftrightarrow [(\sim \varphi) \wedge (\sim \psi)]$$

and

$$\sim (\varphi \wedge \psi) \Leftrightarrow [(\sim \varphi) \vee (\sim \psi)]$$

are called the ***De Morgan laws*** (the second one is also proved by comparing truth tables).

***Implication***: The formula "if $\varphi$, then $\psi$" is denoted by $\varphi \Rightarrow \psi$; here is its truth table.

| $\varphi$ | $T$ | $T$ | $F$ | $F$ |
|---|---|---|---|---|
| $\psi$ | $T$ | $F$ | $T$ | $F$ |
| $\varphi \Rightarrow \psi$ | $T$ | $F$ | $T$ | $T$ |

This definition requires some explanation. Everyone believes that $\varphi \Rightarrow \psi$ should be true whenever $\psi$ is true, and that it should be false when $\varphi$ is true and $\psi$ is false. But why should $\varphi \Rightarrow \psi$ be true when both $\varphi$ and $\psi$ are false? (Remember that a statement must be either true or false once truth values are assigned to its sentence symbols.) As we said at the outset, we use familiar interpretations to guide us to reasonable definitions. Recall that the ***contrapositive*** of the formula $\varphi \Rightarrow \psi$ is

$$(\sim \psi) \Rightarrow (\sim \varphi).$$

Looking at elementary examples in ordinary language indicates that an implication and its contrapositive should be logically equivalent. Consider, for example, the implication, "If $\pi$ is rational, then $\pi^2$ is rational." The contrapositive is "If $\pi^2$ is irrational, then $\pi$ is irrational." This latter formula is correct; after all, were the conclusion false, then $\pi$ would be rational; hence the original implication gives $\pi^2$ rational, contradicting the present hypothesis that $\pi^2$ is irrational. This is the way we proved, in Theorem 4.35, that $\pi$ is irrational. In this informal discussion, we have used the "believable" portion of implication's truth table. Let us now look at the part of the definition that $\varphi \Rightarrow \psi$ is true in case both $\varphi$ and $\psi$ are false. We want $\varphi \Rightarrow \psi$ to have the same truth table as its contrapositive $(\sim \psi) \Rightarrow (\sim \varphi)$. But if both $\varphi$ and $\psi$ are false, then both $(\sim \psi)$ and $(\sim \varphi)$ are true, and so $(\sim \psi) \Rightarrow (\sim \varphi)$ is true.

The formula $\psi \Rightarrow \varphi$ is called the ***converse*** of $\varphi \Rightarrow \psi$. The converse of a true formula need not be true. For example, the converse of "If $f(x)$ is a polynomial, then $f(x)$ is continuous" is the false formula "If $f(x)$ is continuous, then $f(x)$ is a polynomial." The reader will check, in Exercise G.3, that $\varphi \Leftrightarrow \psi$ and $(\varphi \Rightarrow \psi) \wedge (\psi \Rightarrow \varphi)$ are logically equivalent, so that $\varphi \Leftrightarrow \psi$ means that both $\varphi \Rightarrow \psi$ and its converse $\psi \Rightarrow \varphi$ are true.

We have finished our description of the propositional calculus; it consists of formulas, each of which is constructed from sentence symbols $P$, $Q$, ...

and connectives (defined by the truth tables), and each formula has a truth value once truth values are assigned to the sentence symbols in it.

The connectives $\sim$, $\wedge$, $\vee$, $\Leftrightarrow$, and $\Rightarrow$ are not independent of one another. For example, the De Morgan laws display a relation between $\sim$, $\vee$, and $\wedge$. Exercise G.4 below shows that one can define $\Rightarrow$ in terms of $\sim$ and $\wedge$, while Exercise G.3 shows $\varphi \Leftrightarrow \psi$ is logically equivalent to $(\varphi \Rightarrow \psi) \wedge (\psi \Rightarrow \varphi)$; that is, $\Leftrightarrow$ can be defined in terms of $\Rightarrow$ and $\wedge$, and, hence, in terms of $\sim$ and $\wedge$. One can define a new connective, called the *Scheffer stroke*, and the standard five connectives can be defined in terms of it (see Exercise G.8).

## *Exercises*

**G.1.** (i) *(Law of the Excluded Middle)* For any formula $\varphi$, prove that $\varphi \vee (\sim \varphi)$ is a tautology.
(ii) *(Contradiction)* For any formula $\varphi$, prove that $\varphi \wedge (\sim \varphi)$ is always false.

**G.2.** Prove that $\sim (\sim \varphi)$ is logically equivalent to $\varphi$; that is, $\sim (\sim \varphi) \Leftrightarrow \varphi$ is a tautology.

**G.3.** Prove that $\varphi \Leftrightarrow \psi$ and $(\varphi \Rightarrow \psi) \wedge (\psi \Rightarrow \varphi)$ are logically equivalent.

**G.4.** Show that $\varphi \Rightarrow \psi$ is logically equivalent to $(\sim \varphi) \vee \psi$.

**G.5.** Show that $\sim (\varphi \Rightarrow \psi)$ is logically equivalent to $\varphi \wedge (\sim \psi)$.

**G.6.** Show that $(\varphi \vee \psi) \wedge [\sim (\varphi \wedge \psi)]$ and $[\varphi \wedge (\sim \psi)] \vee [\psi \wedge (\sim \varphi)]$ are logically equivalent.

**G.7.** Show that $\varphi \Leftrightarrow \psi$ and $\sim [(\varphi \wedge (\sim \psi)) \vee (\psi \wedge (\sim \varphi))]$ are logically equivalent.

**G.8.** Given formulas $\varphi$ and $\psi$, define the *Scheffer stroke* $\varphi \downarrow \psi$ by the truth table

| $\varphi$ | T | T | F | F |
|---|---|---|---|---|
| $\psi$ | T | F | T | F |
| $\varphi \downarrow \psi$ | F | F | F | T |

(i) Show that $\varphi \downarrow \psi$ is logically equivalent to $(\sim \varphi) \wedge (\sim \psi)$. (One pronounces $\varphi \downarrow \psi$ as "neither $\varphi$ nor $\psi$" because of this result.)
(ii) Show that $\sim \varphi$ is logically equivalent to $\varphi \downarrow \varphi$.
(iii) Show that $\varphi \wedge \psi$ is logically equivalent to $(\sim \varphi) \downarrow (\sim \psi)$.
(iv) Show that $\varphi \vee \psi$ is logically equivalent to $\sim (\varphi \downarrow \psi)$.

**G.9. (Associativity).** For any formulas $\varphi$, $\psi$, and $\theta$, prove that

$$\varphi \wedge (\psi \vee \theta) \Leftrightarrow (\varphi \wedge \psi) \vee \theta \quad \text{and} \quad \varphi \wedge (\psi \wedge \theta) \Leftrightarrow (\varphi \wedge \psi) \wedge \theta$$

are tautologies. (Hint: The truth tables will now begin

| $\varphi$ | $T$ | $T$ | $T$ | $T$ | $F$ | $F$ | $F$ | $F$ |
|-----------|-----|-----|-----|-----|-----|-----|-----|-----|
| $\psi$ | $T$ | $T$ | $F$ | $F$ | $T$ | $T$ | $F$ | $F$ |
| $\theta$ | $T$ | $F$ | $T$ | $F$ | $T$ | $F$ | $T$ | $F$ .) |

**G.10. (Distributivity)** Let $\varphi$, $\psi$, and $\theta$ be formulas.
(i) Prove that $\varphi \vee (\psi \wedge \theta) \Leftrightarrow (\varphi \vee \psi) \wedge (\varphi \vee \theta)$ is a tautology.
(ii) Prove that $\varphi \wedge (\psi \vee \theta) \Leftrightarrow (\varphi \wedge \psi) \vee (\varphi \wedge \theta)$ is a tautology.

## PROOFS

There is a notion of proof in the propositional calculus. Every deductive language involves certain formulas, called *axioms*, and ways to deduce *theorems* (called *proofs*). Axioms are theorems, by definition. The usual deductive rule is called *modus ponens* (literally: mood that affirms—early logicians classified different types of syllogisms into "moods"): if both $\varphi$ and $\varphi \Rightarrow \psi$ are theorems, then $\psi$ is a theorem. A *formal proof* of a formula $\varphi = S_n$ is a finite list of formulas $S_1, S_2, \cdots , S_n$ such that each $S_i$ is either an axiom or follows from two earlier formulas in the list by modus ponens. A formal proof is usually very long and very dull. Now the language of mathematics is more complicated than the propositional calculus. One can set up axioms for mathematics (the most popular today are called the *Zermelo-Fraenkel axioms)*, but a formal proof of a formula as simple as $2 + 1 = 1 + 2$ occupies pages. (It is a good idea, once in your life, to go into a secluded room and actually write a formal proof [see Exercise G.12 at the end of the next section]. This experience should cure you from ever wanting to write another one.) Consequently, a mathematician's proof is not a formal one. Instead, it is an argument designed by and for people (we can view a formal proof as a proof designed for machines). The highlights of a formal proof are given, but not all its steps; the underlying agreement is that one could, if challenged, supply more details; indeed, if necessary, one could even produce a formal proof.

The *law of the excluded middle*, Exercise G.1, is the formal statement of our initial decision that every formula is either true or false; there is no "middle ground." We may regard this "law" as a technique of proof, called *indirect proof* or *proof by contradiction*: a formula $\varphi$ is true if its negation $\sim \varphi$ is false. For example, (a special case of) the *Brouwer fixed point theorem* states that if $f$ is a continuous function from the closed interval $[0, 1]$ to itself, then there exists a *fixed point*; i.e., there is some $a$ in $[0, 1]$ with $f(a) = a$. This theorem is proved indirectly: the assumption that no fixed point exists leads to a contradiction, and one concludes, therefore, that a fixed point must exist. Some people (not me) are uncomfortable with proving existence in this way, but virtually all mathematicians do accept this technique. (One can imagine how such niceties might influence theological arguments; of course, theological arguments involve imprecisely defined terms and occur within richer languages than the propositional calculus.)

To this point, Aristotle would feel right at home with our discussion; he would not be familiar with our notation, but the ideas they represent are essentially in his writings. But here are some questions and results that would have astounded him. One can ask whether every true formula in a language is provable; that is, given a tautology $\varphi$, is there a formal proof of $\varphi$? For some languages, e.g., the propositional calculus, the answer is "yes": true and provable are the same thing. This result, the so-called *Completeness Theorem*, was proved by K. Gödel in 1930; but, in 1931, he proved the *Incompleteness Theorem*: If a language is rich enough to express ordinary arithmetic, then there are always *undecideable* formulas $\varphi$ (which are true in natural interpretations) such that neither $\varphi$ nor $\sim \varphi$ can be proved in the language.

### Exercises

**G.11.** (i) Define a sequence $r_1, r_2, \ldots, r_n, \ldots$, where $r_n$ is the world record for the fastest mile ever run in or before year $n$. Show that this sequence converges. (Hint: One says that a sequence $\{a_n\}$ is *bounded from below* if there exists a number $B$ with $B \leq a_n$ for all $n$. Use the fact that a decreasing sequence bounded from below converges.)

(ii) Conclude that there is a time $t$ so that runners can run a mile in times arbitrarily close to $t$, yet no runner will ever run a mile even a picosecond faster than $t$. (Hint: Use Exercise 3.32.) There are some who balk at our asserting the existence of a number no one will ever know how to compute.

## SET THEORY

We now describe some set theory in an informal way. A **set** $X$ is a collection of elements, and two sets $X$ and $Y$ are called **equal**, denoted by

$$X = Y,$$

if they are comprised of precisely the same elements. We write

$$x \in X$$

to denote "$x$ is an **element** of $X$" (if $x \in X$, one also says that "$x$ **belongs** to $X$" or that "$x$ is a **member** of $X$"). We say that a set $X$ is a **subset** of a set $Y$, denoted by

$$X \subset Y,$$

if every element of $X$ is also an element of $Y$: formally, if $x \in X$, then $x \in Y$ (if $X \subset Y$, one also says that $X$ is **included** in $Y$ or that $X$ is **contained** in $Y$). Note that $X = Y$ if and only if $X \subset Y$ and $Y \subset X$. This last remark underlies many proofs in which one wants to prove that two sets are equal. For example, let

$$X = \{ \text{ all real numbers } r \text{ with } r \geq 0 \}$$

and let

$$Y = \{ \text{ all real numbers } s \text{ with } s = a^2 \text{ for some } a \}.$$

The proof that $X = Y$ involves two steps: $X \subset Y$ and $Y \subset X$. To see that $X \subset Y$, let $r \in X$. Since $r \geq 0$, it has a real square root $a = \sqrt{r}$, and $r = a^2$; therefore $r \in Y$. For the reverse inclusion, let $s = a^2 \in Y$. If $a \geq 0$, then $s \geq 0$, as desired; if $a < 0$, then $a = -b$ for some positive number $b$, so that $s = a^2 = (-b)^2 = (-1)^2 b^2 = b^2 > 0$ [because $(-1)^2 = 1$, as we proved in Chapter 4]. In either case, $s \in X$, as desired. We conclude that $X = Y$.

Notice that $X \subset X$ is always true. If $X$ is a **proper** subset of $Y$, that is, if $X \subset Y$ and $X \neq Y$, we write $X \subsetneq Y$. Impressed by the analogy between inclusion of subsets and the inequality symbols $\leq$ and $<$ for real numbers, some authors denote inclusion by $\subseteq$ and strict inclusion by $\subset$. I do not think that this is a good idea, for it is better to use the simpler notation ($\subset$ instead of $\subseteq$) for the circumstance occurring most frequently.

Here are some ways to make new subsets out of old ones.

**Complement**: If $X$ is a subset of a set $U$, its **complement** is

$$\sim X = \{u \in U : u \notin X\};$$

that is, the complement of $X$ consists of all those elements of $U$ that do not belong to $X$.

**Union**: If $X$ and $Y$ are subsets of a set $U$, then their **union**, denoted by $X \cup Y$, consists of all $u$ with $u \in X$ or $u \in Y$.

**Intersection**: If $X$ and $Y$ are subsets of a set $U$, then their **intersection**, denoted by $X \cap Y$, consists of all $u$ with $u \in X$ and $u \in Y$.

In order that the intersection of two subsets always be defined, we need the **empty set** $\emptyset$. By definition, $\emptyset$ is the subset having no elements (after all, two subsets may have no elements in common). In Exercise G.22, you will show that $\emptyset$ is a subset of every set.

One can picture $X \cap Y$ and $X \cup Y$ in **Venn diagrams** [named after J. Venn (1834–1923)].

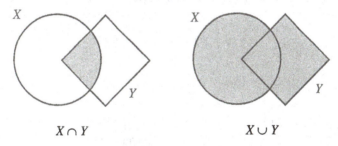

$$X \cap Y \qquad\qquad\qquad X \cup Y$$

Figure G.1

There is a strong analogy between formulas of propositional calculus and Venn diagrams. This should not be a surprise, for union is defined in terms of *or*, intersection is defined in terms of *and*, and complement is defined in terms of *not*. We are going to describe a way of regarding formulas as subsets in such a way that the connectives correspond to the set operations.

There is a language of arithmetic, much richer than that of the propositional calculus. In particular, it involves the notion of **variables**. One can define variables precisely, but we will treat them in an informal manner: each variable is to be regarded as a pronoun that can be assigned a particular range of values. For example, if we say that $n$ is a positive integer, then the formula $n + 5 > 26$ yields a statement for each choice of $n = 1, 2, 3, \ldots$. We will write $\varphi = \varphi(x)$ to denote a formula involving a variable $x$. If $\varphi = \varphi(x)$, where $x$ can be assigned values in a set $X$, define the subset $V(\varphi)$ of $X$ by

$$V(\varphi) = \{k \in X : \varphi(k) \text{ is true}\}$$

(we are assuming, for each $k \in X$, that the formula $\varphi(k)$ has a truth value). For example, if $n$ is a positive integer variable and $\varphi(n)$ is the statement "$n$ has remainder 1 after dividing by 3." then

$$V(\varphi) = \{1, 4, 7, \ldots\}.$$

The connectives can be defined for formulas with variables, so that $\sim \varphi(x)$, $\varphi(x) \wedge \psi(x)$, $\varphi(x) \vee \psi(x)$, $\varphi(x) \Leftrightarrow \psi(x)$, and $\varphi(x) \Rightarrow \psi(x)$ make sense. For example, $\sim \varphi(x)$ is true means that $\sim \varphi(k)$ is true for every $k$ in $X$. The reader can now see that the connectives correspond to the subset operations:

$$V(\sim \varphi) \text{ is the complement} \sim V(\varphi);$$

$$V(\varphi \wedge \psi) = V(\varphi) \cap V(\psi);$$

$$V(\psi \vee \psi) = V(\varphi) \cup V(\psi).$$

We shall consider $V(\varphi \Rightarrow \psi)$ and $V(\varphi \Leftrightarrow \psi)$ in Exercises G.14 and G.15.

The following observation is fundamental: Given $\varphi(x)$ and $\psi(x)$, if $\varphi(x) \Leftrightarrow \psi(x)$ is true for every $x$, then $V(\varphi) = V(\psi)$. One sees this by showing that $V(\varphi) \subset V(\psi)$ and $V(\psi) \subset V(\varphi)$. For the first inclusion, take $a \in V(\varphi)$. By definition of $V(\varphi)$, the formula $\varphi(a)$ is true. Because $\varphi(a) \Leftrightarrow \psi(a)$, the formula $\psi(a)$ is true, and so $a \in V(\psi)$. Therefore, $V(\varphi) \subset V(\psi)$. The reverse inclusion $V(\psi) \subset V(\varphi)$ is proved in the same way, and so $V(\varphi) = V(\psi)$.

The next circle of ideas in the study of set theory involves relations and functions, and the next idea afterwards involves counting and the study of infinite numbers. A second direction one can go involves Boolean algebra, a common generalization of propositional calculus and set theory.

## *Exercises*

**G.12.** If $a$ and $b$ are numbers, define the *ordered pair* $(a, b)$ by

$$(a, b) = \{a, \{a, b\}\};$$

that is, $(a, b)$ is a set having exactly two elements, namely, the number $a$ and the set $\{a, b\}$ (if $a = b$, then $\{a, b\} = \{a\}$, the set having $a$ as its only element; otherwise, $\{a, b\}$ is a 2-element set).

(i) Prove that $(a, b) = (a', b')$ if and only if $a = a'$ and $b = b'$. (You may assume one of the Zermelo-Fraenkel axioms, the *axiom of foundation*, which implies that

$$x \in y \in x$$

is always false.)

(ii) Give a formal proof that $(a, b) = (a', b')$ if and only if $a = a'$ and $b = b'$. (This problem is not well posed because we have not given a list of axioms, but the spirit of this exercise is that a sizable fragment of a formal proof should be written.)

(iii) Here is an alternative definition of ordered pair:

$$[a, b] = \{\{a\}, \{a, b\}\};$$

that is, $[a, b]$ is the set whose elements are the 1-element set $\{a\}$ and, if $a \neq b$, the 2-element set $\{a, b\}$ (if $a = b$, then $\{a, b\} = \{a\}$, in which case $\{\{a\}, \{a\}\}$ is the set whose only element is $\{a\}$).

Prove that $[a, b] = [a', b']$ if and only if $a = a'$ and $b = b'$.

**G.13.** For any subset $X$ of a set $U$, prove that $X \cup (\sim X) = U$ and $X \cap (\sim X) = \emptyset$.

**G.14.** Define $X - Y = \{u \in X$ and $u \notin Y\}$. If $\varphi = \varphi(u)$ and $\psi = \psi(u)$ are formulas, show that $V(\varphi \Rightarrow \psi) = {} \sim [V(\varphi) - V(\psi)]$.

**G.15.** If $X$ and $Y$ are subsets, one defines

$$X \triangle Y = (X - Y) \cup (Y - X)$$

and calls it their *symmetric difference*. Show that

$$V(\varphi \Leftrightarrow \psi) = {} \sim [V(\varphi) \triangle V(\psi)].$$

**G.16.** Prove that $V(\varphi \downarrow \psi) = [\sim V(\varphi)] \cap [\sim V(\psi)]$, where $\varphi = \varphi(x)$ and $\psi = \psi(x)$. (See Exercise G.8.) (This is the same as $\sim [V(\varphi) \cup V(\psi)]$, by De Morgan.)

**G.17.** *(Associativity).* For any subsets $X$, $Y$, and $Z$ of a set $U$, prove that

$$X \cup (Y \cup Z) = (X \cup Y) \cup Z \quad \text{and} \quad X \cap (Y \cap Z) = (X \cap Y) \cap Z.$$

**G.18.** *(Distributivity)* For any subsets $X$, $Y$, and $Z$ of a set $U$, prove that

$$X \cup (Y \cap Z) = (X \cup Y) \cap (X \cup Z)$$

and

$$X \cap (Y \cup Z) = (X \cap Y) \cup (X \cap Z).$$

## QUANTIFIERS

The most interesting statements in mathematics involve variables: Every even integer $\geq 4$ is a sum of two primes; If $n \geq 3$, there do not exist positive integers $a$, $b$, $c$ with $a^n + b^n = c^n$; For every $\varepsilon > 0$ there exists an integer $\ell$ such that $|a_n - L| < \varepsilon$ whenever $n \geq \ell$. The following notation is now quite common. If $x$ is a variable varying over a set $X$, and if $\varphi(x)$ is a formula, then the statement "$\varphi(x)$ is true for all $x \in X$" is written

$$(\forall x \in X)[\varphi(x)];$$

one calls $\forall$ the ***universal quantifier***; when the range $X$ is known, one usually abbreviates this as $(\forall x)[\varphi(x)]$. If we recall Venn diagrams, we are saying that $(\forall x)[\varphi(x)]$ is true if $V([\varphi(x)]) = X$. For example, if $S(0)$, $S(1)$, $S(2)$, ..., $S(n)$, ... is a sequence of formulas and $\mathbb{N}$ denotes the set of all nonnegative integers, then one can write the inductive step as

$$(\forall n \in \mathbb{N})[S(n) \Rightarrow S(n+1)].$$

Another kind of statement is "There exists an $x \in X$ for which $\varphi(x)$ is true." The common notation is:

$$(\exists x \in X)[\varphi(x)];$$

one calls $\exists$ the ***existential quantifier***; when the range $X$ is known, one usually abbreviates this as $(\exists x)[\varphi(x)]$. We are saying that $(\exists x)[\varphi(x)]$ is true if $V([\varphi(x)] \neq \emptyset$.

The statement of Fermat's Last Theorem involves two quantifiers. In symbols, it says:

$$(\forall n \geq 3) \sim (\exists \text{ positive integers } a, b, c)[a^n + b^n = c^n].$$

That a number $L$ is the limit of a sequence $\{a_n\}$ has three quantifiers.

$$(\forall \varepsilon > 0)(\exists \text{ positive integer } \ell)(\forall n \geq \ell)[|a_n - L| < \varepsilon].$$

Here are two basic facts involving quantifiers and negation:

$$\sim (\forall x)[\varphi(x)] \Leftrightarrow (\exists x)[\sim \varphi(x)];$$

and

$$\sim (\exists x)[\varphi(x)] \Leftrightarrow (\forall x)[\sim \varphi(x)].$$

The reason for the first statement is quite clear: to say that $\varphi(x)$ is not true for all $x$ merely says that there is at least one $x$ for which $\varphi(x)$ is false; i.e., there is at least one $x$ for which $\sim \varphi(x)$ is true. The second statement is just the negation of the first one.

One can negate formulas involving several quantifiers by applying this basic fact several times. We illustrate by stating the results for two quantifiers. Let $\varphi(x, y)$ be a formula involving two variables.

$$\sim (\forall x)(\forall y)[\varphi(x, y)] \Leftrightarrow (\exists x)(\exists y)[\sim \varphi(x, y)].$$

$$\sim (\forall x)(\exists y)[\varphi, (x, y)] \Leftrightarrow (\exists x)(\forall y)[\sim \varphi(x, y)].$$

$$\sim (\exists x)(\forall y)\varphi(x, y)] \Leftrightarrow (\forall x)(\exists y)[\sim \varphi(x, y)].$$

$$\sim (\exists x)(\exists y)[\varphi(x, y)] \Leftrightarrow (\forall x)(\forall y)[\sim \varphi(x, y)].$$

The proof of the first logical equivalence is clear:

$$\sim (\forall x)(\forall y)[\varphi(x, y)] \Leftrightarrow (\exists x) \sim ((\forall y)[\varphi(x, y)])$$

$$\Leftrightarrow (\exists x)(\exists y)[\sim \varphi(x, y)].$$

The proofs of the other logical equivalences are similar.

Here is a practical example using this formalism. Suppose we want to prove that the sequence $\{a_n\}$ with $a_n = (-1)^n$ for all $n \geq 0$ does not converge to 0. Write $a_n \to 0$ in a very formal way:

$$(\forall \varepsilon > 0)(\exists \ell)(\forall n \geq \ell)[|(-1)^n - 0| < \varepsilon];$$

its negation is:

$$(\exists \varepsilon > 0)(\forall \ell)(\exists n \geq \ell)[|(-1)^n - 0| \geq \varepsilon].$$

Now $|(-1)^n - 0| = 1$ for all $n$, and so we are asked to show that there is some positive number $\varepsilon$ so that $1 \geq \varepsilon$. This is easy: take $\varepsilon = \frac{1}{2}$. We conclude that $(-1)^n \not\to 0$. Some formulas are rather complicated, and so setting them up in such a formal way allows them to be negated mechanically; see Exercise G.21.

Note that the order of the quantifiers makes a difference. Consider the formulas (where $\mathbb{Z}$ is the set of integers and $\mathbb{R}$ is the set of real numbers):

$$(\forall x \in \mathbb{R})(\exists k \in \mathbb{Z})[x < k];$$

$$(\exists k \in \mathbb{Z})(\forall x \in \mathbb{R})[x < k].$$

In words, the first formula says that for any given number $x$, there is some integer $k$ larger than it. Of course, this statement is true: for example, if $x = e = 2.71828\cdots$, then $k = 3$ "works"; if $x = 100\pi$, then $k = 400$ works; and so forth. Notice that the integer $k$ that works depends on the particular choice of $x$; for example, $k = 3$ does not work for $100\pi$. The second formula says that there is some integer $k_0$ which is larger than every number $x$; that is, $k_0$ works simultaneously for every $x$. Of course, this second formula is false; there is no largest number. Here is a familiar example. A clothing store has socks to fit any customer, but there are different sized socks for different sized folks. "Stretch socks" are advertised as "one size fits all"; one pair of stretch socks can fit anyone.

A second basic fact about quantifiers is this: If $\varphi(x, k)$ is a formula, where $X$ and $K$ are sets, then

$$(\exists k \in K)(\forall x \in X)[\varphi(x, k)] \text{ implies } (\forall x \in X)(\exists k \in K)[\varphi(x, k)].$$

That this should be so is quite reasonable. If some $k_0$ works for all $x$, then there is a single $k$ working for every particular $x$ (namely, $k = k_0$). If one size fits all, then everyone can be fitted.

There are much better reasons for the quantifier formalism than its allowing one to negate complicated sentences. The study of languages with quantifiers is needed, for example, to prove Gödel's incompleteness theorem; it is also needed to understand how computers work.

## *Exercises*

**G.19.** Prove that the sequence $\{(-1)^n\}$ does not converge to 1.

**G.20.** For each $n \geq 1$, let $f_n(x)$ be a real valued function defined on a closed interval $[a, b]$. If $L(x)$ is another such function, then one says that $f_n(x)$ *converges pointwise* to $L(x)$ on $[a, b]$ if $f_n(c) \to L(c)$ for each $c \in [a, b]$.
(i) If $f_n(x) = x^n$, prove that $f_n(x)$ converges pointwise to $L(x)$ on $[0, 1]$, where $L(x) = 0$ if $0 \leq x < 1$ and $L(1) = 1$.
(ii) Write the definition of pointwise convergence with quantifiers, and then write its negation.

**G.21.** For each $n \geq 1$, let $\{f_n(x)\}$ be a sequence of real valued functions defined on a closed interval $[a, b]$. If $L(x)$ is another such function, then one says that $f_n(x)$ *converges uniformly* to $L(x)$ on $[a, b]$ if, for each $\varepsilon > 0$, there is an integer $\ell$ so that $|f_n(c) - L(c)| < \varepsilon$ for all $c \in [a, b]$ and all $n \geq \ell$.
(i) Write the definition of uniform convergence with quantifiers.
(ii) Show that if $f_n(x)$ converges uniformly to $L(x)$ on $[a, b]$, then $f_n(x)$ converges pointwise to $L(x)$ on $[a, b]$.
(iii) Write the negation of uniform convergence.
(iv) If $f_n(x) = x^n$, prove that $f_n(x)$ does not converge uniformly to $L(x)$ on $[0, 1]$, where $L(x) = 0$ if $0 \leq x < 1$ and $L(1) = 1$.

**G.22.** (i) Prove that the empty set $\emptyset$ is a subset of every set $X$. (Hint: The negation of $(\forall x)[x \in \emptyset \Rightarrow x \in X]$ is $(\exists x)[x \in \emptyset$ and $x \notin X]$; but $x \in \emptyset$ is false.)
(ii) Prove that there is only one empty set.

# Index